SpringerBriefs in Mathematics

SpringerBriefs present concise summaries of cutting-edge research and practical applications across a wide spectrum of fields. Featuring compact volumes of 50 to 125 pages, the series covers a range of content from professional to academic. Briefs are characterized by fast, global electronic dissemination, standard publishing contracts, standardized manuscript preparation and formatting guidelines, and expedited production schedules.

Typical topics might include:

A timely report of state-of-the art techniques A bridge between new research results, as published in journal articles, and a contextual literature review A snapshot of a hot or emerging topic An in-depth case study A presentation of core concepts that students must understand in order to make independent contributions

SpringerBriefs in Mathematics showcases expositions in all areas of mathematics and applied mathematics. Manuscripts presenting new results or a single new result in a classical field, new field, or an emerging topic, applications, or bridges between new results and already published works, are encouraged. The series is intended for mathematicians and applied mathematicians. All works are peer-reviewed to meet the highest standards of scientific literature.

Titles from this series are indexed by Scopus, Web of Science, Mathematical Reviews, and zbMATH.

Seiro Omata • Karel Svadlenka • Elliott Ginder

Variational Approach to Hyperbolic Free Boundary Problems

 Springer

Seiro Omata
Kanazawa University
Kanazawa, Japan

Karel Svadlenka
Kyoto University
Kyoto, Japan

Elliott Ginder
Meiji University
Tokyo, Japan

ISSN 2191-8198   ISSN 2191-8201   (electronic)
SpringerBriefs in Mathematics
ISBN 978-981-19-6730-6   ISBN 978-981-19-6731-3   (eBook)
https://doi.org/10.1007/978-981-19-6731-3

This Springer imprint is published by the registered company Springer Nature Singapore Pte Ltd.
The registered company address is: 152 Beach Road, #21-01/04 Gateway East, Singapore 189721, Singapore

Preface

This volume deals with free boundary problems for partial differential equations of hyperbolic type. The focus is on hyperbolic problems with variational structure. Then a weak solution can be constructed as a limit of approximate solutions, which are in turn obtained as time-interpolated minimizers of variational functionals with space variables kept continuous and time variable discretized. We attempt to summarize all that is known to date about this analytical approach. In the second part of the book, we pursue extensions to more complicated, in particular vector-valued, problems. For example, we discuss the problem of a rotating elastic body contacting a solid obstacle, the problem of a bouncing elastic shell, or the oscillatory motion by mean curvature of a network of interfaces with junctions.

Kanazawa, Japan

Kyoto, Japan

Tokyo, Japan

July 2022

Seiro Omata

Karel Svadlenka

Elliott Ginder

Contents

Symbols

$C^k(D)$	space of functions with continuous k-th derivatives
$C^{0,\alpha}(D)$	space of Hölder continuous functions
$C_0^\infty(D)$	space of infinitely differentiable functions with a compact support in D
$C([0, T]; V)$	the Banach space of continuous mappings $[0, T] \to V$
$\mathscr{H}^n$	n-dimensional Hausdorff measure
$H^k(D)$	the Sobolev space $W^{k,2}(D)$
$H_0^k(D)$	the subspace of $H^k(D)$ with zero trace on ∂D
$\|\cdot\|_{H^k(D)}$	norm of Sobolev space $H^k(D)$
$H^1(0, T; V)$	the Sobolev–Bochner space $W^{1,2}(0, T; V)$
$\mathscr{L}^n$	n-dimensional Lebesgue measure
$|D|$	Lebesgue measure of a set D
$L^p(D)$	the Lebesgue space of functions with integrable p-powers for $1 \le p < \infty$
$L^\infty(D)$	the Lebesgue space of essentially bounded functions
$L_{loc}^p(D)$	the Lebesgue space of functions with locally integrable p-powers
$\|\cdot\|_{L^p(D)}$	norm of Lebesgue space $L^p(D)$
$L^p(0, T; V)$	the space of Bochner measurable p-integrable functions $(0, T) \to V$
$W^{k,p}(D)$	the Sobolev space of functions with all k-th derivatives in $L^p(D)$
$W^{k,p}(0, T; V)$	the Sobolev–Bochner space of functions $(0, T) \to V$
χ_A	the characteristic function of a set A
∂D	the boundary of a set D
$D\eta$	the Jacobian matrix of a function η
$\langle \cdot, \cdot \rangle$	the inner product in $L^2(D)$
$\{u > 0\}$	set of all points z such that $u(z) > 0$
$\to$	strong convergence in a normed linear space
$\rightharpoonup$	weak convergence in a normed linear space
$\rightrightarrows$	uniform convergence
spt f	support of a continuous function f
$A \backslash B$	set difference of sets A and B
$D' \Subset D$	set D' is compactly included in a set D

Chapter 1
Introduction

This volume is devoted to the study of hyperbolic free boundary problems. Such problems can be used to model, among others, the motion of a droplet on a solid obstacle, the motion of a bubble on a water surface or the vibration of a string hitting an obstacle. For the purpose of explanation, we restrict ourselves to the scalar case, where the membrane surrounding the water or the air in the case of a droplet or a bubble can be expressed as the graph of a function. However, since the mathematical analysis of the interaction between the membrane and the obstacle can be extended to the vector-valued case, we shall later on present specific examples investigating such problems.

In general, the membrane forms a positive contact angle with the obstacle, and therefore the Laplacian or other differential operator describing the shape of the membrane is only a measure supported in the boundary of the contact set, usually called *a free boundary*. We will show how to derive mathematical problems for a few physical systems starting from an action functional. Then we will discuss the mathematical theory, and introduce approximation methods for obtaining numerical solutions of the model equations.

Such problems were first investigated by Ta-Tsien Li and his collaborators (see, e.g., [43–46]), where a local, and later global, existence theory for a class of one-dimensional quasilinear hyperbolic free boundary problems was established. Although the study was successful in obtaining rather general existence results, we remark that the motivation of their research is to understand problems such as shock behavior in the generalized Riemann problem for isentropic flow. This viewpoint specifies the problem setting and the underlying assumption for which their theory is applicable.

The current study focuses on a different class of hyperbolic free boundary problems: those which are characterized by the presence of a variational structure. By discretizing time, our hyperbolic problems can then be recast as sequences of minimization problems, and this enables us to employ modern techniques from the calculus of variations. An advantage of this approach is that its framework can be extended to more general problems in a natural way. In particular, fully nonlinear problems, and problems with constraints, fall within the scope of our method. Another important aspect is that it develops existence theory constructively, and thus translates directly into numerical approximation schemes.

Although the existence theory that we establish in this book is limited to the one-dimensional setting, similarly to [43], we note that uniformly bounded approximate solutions can be obtained in arbitrary dimensions. On the other hand, in the setting of transmission problems with free boundary, i.e., problems similar to those considered by Li [43], the proof of stability of multidimensional shocks is a well-known work of A. Majda [48–50] (but see also [51] for a different type of higher-dimensional hyperbolic free boundary problem related to the atmospheric sciences).

Investigation of hyperbolic free boundary problems in the variational framework was introduced relatively recently. As one of the first examples, K. Kikuchi and S. Omata considered the problem of removing an adhesive tape from a solid plane [35]. In this phenomenon, when the tape is lifted up at its edge, it may peel off from the plane while vibrating. It should be noted that the boundary between the region where the tape is attached to the plane, and the region where it is detached, changes with time and is, therefore, a free boundary. For this problem, in one spatial dimension, K. Kikuchi and S. Omata showed, under certain compatibility conditions, the unique existence of a strong solution. In particular, the problem is well-posed. Later, T. Shinohara and K. Nakane showed the existence of a global solution in time [61] and of a periodic solution [62], while H. Imai, K. Nakane, K. Kikuchi and S. Omata [28] developed a numerical method for the one-dimensional problem.

To make the model more physically relevant, constrained problems were also considered, for example, the motion of an incompressible water droplet. The problem then includes global information, manifested as an outer force, and the analysis at the level of differential (momentum) equations becomes cumbersome. As a remedy, a variational approach, based on an action integral, was introduced and the problem was reformulated as looking for a stationary point of the action under the given constraint. To take advantage of this formulation, we use the approach due to N. Kikuchi [33] and A. Tachikawa [86], the so-called *discrete Morse flow* for hyperbolic equations. This method approximates the stationary point by an iterative sequence of minimizations, and can serve as a basis for numerical methods. The approach was exploited in several papers [3, 26, 32, 34, 60, 65, 66] by the Japanese research group of S. Omata and others [59], and further extended to semilinear and nonlocal operators [11, 12] by the Italian research group of G. Orlandi. Finally, an improved scheme which guarantees energy preservation was proposed in [3].

Chapter 2
Physical Motivation

In this chapter, we present several examples of phenomena that can be modeled by hyperbolic free boundary problems, and illustrate some of their distinctive features: collision of a membrane with a rigid obstacle, peeling of an adhesive tape off a solid surface, and motion of a liquid droplet on a solid substrate [67].

2.1 Membrane Collision with an Obstacle

Here we consider a simple model describing the collision of a rubber film or membrane with an obstacle. This is related to the bouncing phenomena which we will discuss later in Chap. 6. We describe the position of the membrane by the graph of a scalar function $u(x, t) : \Omega \times [0, \infty) \to \mathbb{R}$, where Ω is a domain in $\mathbb{R}^m$. The obstacle is located at the zero level set of u, and perfectly inelastic collisions are assumed, i.e., the local reflection rate between the membrane and the obstacle is zero. The case of fully elastic reflection is treated in the series of papers [7, 71, 78–80] and in [36] where, unlike usual bouncing phenomena, energy is conserved and the film hits the obstacle many times. This necessarily leads to a different mathematical approach, namely the penalty method. We also mention the theories developed in [6] for one-point obstacles and in [47] for systems with finite degrees of freedom.

The initial-boundary value problem for our obstacle problem with zero local reflection rate is to find a function $u : \Omega \times [0, T) \to \mathbb{R}$ satisfying

$$
\begin{cases}
\chi_{\overline{\{u>0\}}} u_{tt} &= \Delta u \quad \text{in} \quad Q_T := \Omega \times (0, T), \\
u(x, 0) &= u_0(x) \quad \text{in} \quad \Omega, \\
u_t(x, 0) &= v_0(x) \quad \text{in} \quad \Omega, \\
u(x, t) &= f(x, t) \quad \text{on} \quad \partial\Omega.
\end{cases}
\tag{2.1}
$$

© The Author(s), under exclusive license to Springer Nature Singapore Pte Ltd. 2022
S. Omata et al., *Variational Approach to Hyperbolic Free Boundary Problems*,
SpringerBriefs in Mathematics, https://doi.org/10.1007/978-981-19-6731-3_2

Fig. 2.1 Common setup of
free boundary problems
presented in Chap. 2

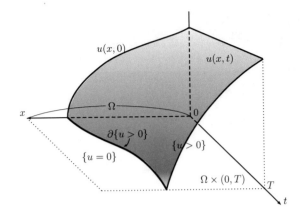

Here, Ω is the spatial domain over which the membrane moves, χ_E denotes the characteristic function of a set E, and $\{u > 0\}$ is a shorthand for the set $\{(x, t) \in \Omega \times (0, T) : u(x, t) > 0\}$ (see Fig. 2.1 for basic notation).

Let us explain the derivation of the above model equation. We consider two scenarios. In the first, the membrane is rising and energy is conserved, while in the second, the membrane hits the obstacle and energy dissipates.

In the energy-preserving case, if we consider the membrane's elastic or tension energy of the form $\int_\Omega |\nabla u|^2 \, dx$, the equation describing the motion of the membrane becomes the wave equation. Indeed, allowing only vertical motion of the membrane gives the kinetic energy $\int_\Omega u_t^2 \, dx$ but, since the membrane stops moving when it touches the obstacle, both kinetic energy and tension energy should only be present in the set $\{u > 0\}$. In other words, we can define the action integral in the form

$$I(u) = \int_0^T \int_\Omega \left((u_t)^2 - |\nabla u|^2\right) \chi_{\{u>0\}} \, dx \, dt. \tag{2.2}$$

Now we calculate the first variation of $I(u)$. We assume that a stationary point u exists and is a continuous function, so that we can select a test function $\zeta \in C_0^\infty(Q_T \cap \{u > 0\})$ whose support is included in $\{u > 0\}$. Then we have $(d/d\varepsilon)I(u + \varepsilon\zeta)|_{\varepsilon=0} = 0$, which yields

$$u_{tt} = \Delta u \quad \text{in} \quad Q_T \cap \{u > 0\} \tag{2.3}$$

in the weak sense, and where Q_T is as in (2.1).

Relatedly, we can also calculate the so-called inner variation, $(d/d\varepsilon)I(u \circ \tau_\varepsilon^{-1})|_{\varepsilon=0} = 0$, with $\tau_\varepsilon = \text{Id} + \varepsilon\eta$, where Id is the identity map and $\eta \in C_0^\infty(Q_T; \mathbb{R}^m \times \mathbb{R})$. After a formal calculation, the details of which will be shown later in Sect. 2.2.2, we have the free boundary condition

$$|\nabla u|^2 - (u_t)^2 = 0 \quad \text{on} \quad Q_T \cap \partial\{u > 0\}. \tag{2.4}$$

On the other hand, in the energy-losing case, i.e., when the membrane hits the obstacle, the above Hamilton's principle cannot be applied. Since the local reflection ratio is 0, when the membrane touches the plane, it instantly comes to a halt and attaches itself. Therefore, it is reasonable to postulate that, when the velocity of the membrane is negative, the positive part max$\{u, 0\}$ of the solution u of the usual wave equation becomes a solution. One way of expressing this is by truncating the acceleration term through its multiplication with the characteristic function of the set $\overline{\{u > 0\}}$. Thus we can write the equation as

$$\chi_{\overline{\{u>0\}}} u_{tt} = \Delta u. \tag{2.5}$$

In the set where the acceleration term vanishes, the equation is just the Laplace equation and therefore we expect that for almost every t the solution is nonnegative. Moreover, the free boundary condition is satisfied if there is no extra force, such as a Radon measure, on the free boundary. A more general calculation supporting this statement will be shown later in Proposition 2.1. Here we state it as a remark.

Remark 2.1 If there is a nonnegative function v which is a classical solution of the equation $v_{tt} - \Delta v = 0$ in $Q_T \cap \{v > 0\}$, and if, moreover, the free boundary $\partial\{v > 0\}$ is smooth and there is no energy concentration on it, i.e., for all $\zeta \in C_0^\infty(Q_T)$,

$$\int_{Q_T} \zeta(v_{tt} - \Delta v) \, dx \, dt = 0, \tag{2.6}$$

then $|\nabla v|^2 - (v_t)^2 = 0$ on $Q_T \cap \partial\{v > 0\}$.

If the membrane is rising, the energy is conserved and the solution satisfies the free boundary condition. Hence the free boundary moves along the characteristic surface of the wave operator. On the other hand, if the graph moves toward and touches the obstacle, the solution is obtained by truncating the solution of wave equation at zero height.

Understanding Eq. (2.5) in this way provides a reasonable description of the phenomenon. Nevertheless, although the equation seems simple, it is not easy to establish its general existence theory. This topic will be discussed in the following chapters.

2.2 Peeling of Adhesive Tape

Here we consider the phenomenon of removing an adhesive tape from a surface. An adhesive tape is uniformly stretched (with a constant tension force S) and attached to a solid plane. Subsequently, the tape is lifted up at its edge, which leads to its gradual peeling off from the plane [13].

The boundary between the attached and detached parts of the tape evolves during the peeling process and becomes a free boundary in this problem. We shall show that this phenomenon can be modeled by an obstacle problem of hyperbolic type with an adhesive force. This leads to a time-dependent contact angle condition, i.e., to a dynamic Young's law. In the stationary case, the contact angle is determined by the difference of the surface tension of the tape between the part that is attached to the surface and the part that is detached. This approach can be generalized and applied to many other problems, including the motion of soap bubble clusters or the motion of droplets on surfaces.

2.2.1 Energy at Equilibrium

Let us calculate the potential energy of a stationary configuration of tape. The ground state is chosen so that the tape is completely attached to the surface. We assume that the substrate is flat, that the tape is relatively thin and that, by choosing a suitable coordinate system, its shape can be described as a hypersurface corresponding to the graph of a scalar function u. The tension force S of the tape is assumed to be always the same, which is true, for example, in the case of a soapy film. Under these assumptions, the energy stored in a stretched tape is given by $\int_{\{u>0\}} S(\sqrt{1 + |\nabla u|^2} - 1)\, dx$.

We also assume that the adhesion force has a potential, i.e., there is an energy associated to the peeling process. As depicted in Fig. 2.2, we can determine the angle θ at the free boundary by investigating the balance between vertical components of the tension force, and the maximum adhesion force $\widetilde{Q}$ in the vertical direction.

The balance condition is $S \sin\theta = \widetilde{Q}$, which implies that the force needed to peel the tape off is $S(1 - \cos\theta) = S(1 - \sqrt{1 - (\widetilde{Q}/S)^2})$. The corresponding work is proportional to the area of the region where the tape is peeled off.

Consequently, the total potential energy within a domain $\Omega \subset \mathbb{R}^m$ can be expressed as

$$\int_{\{u>0\}\cap\Omega} S\left(\left(\sqrt{1 + |\nabla u|^2} - 1\right) + \left(1 - \sqrt{1 - (\widetilde{Q}/S)^2}\right)\right) dx$$

$$= S\int_\Omega \left(\sqrt{1 + |\nabla u|^2} + \left(1 - \sqrt{1 - (\widetilde{Q}/S)^2}\right)\chi_{\{u>0\}}\right) dx - S|\Omega|,$$

Fig. 2.2 Force balance in the model of tape peeling

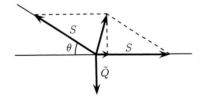

and the stationary shape of the tape is given by the minimizer of the above functional [89]. Note that the same form of potential energy is obtained when the adhesion energy is assumed to be proportional to the area of the attached part of the tape.

If we assume that the contact angle θ is small, we can approximate the energy by the Dirichlet integral

$$L_{\text{adh}}(u) := \frac{S}{2} \int_{\Omega} \left(|\nabla u|^2 + Q^2 \chi_{\{u>0\}} \right) dx, \tag{2.7}$$

which was introduced by H.W. Alt, L.A. Caffarelli and A. Friedman [4, 5], and further studied by S. Omata and Y. Yamaura [64, 68, 69]. Here, we have defined $Q^2 = 2\left(1 - \sqrt{1 - (\widetilde{Q}/S)^2}\right)$.

In [4], it was shown that a minimizer of (2.7) exists in $\{u \in L^1_{\text{loc}}(\Omega); \nabla u \in L^2(\Omega), u = u_0$ on $\partial\Omega\}$, and Δu is a Radon measure whose support is on $\partial\{u > 0\}$, and which is absolutely continuous with respect to the $(m - 1)$-dimensional Hausdorff measure. The authors also discuss the regularity of the free boundary $\partial\{u > 0\}$. Moreover, any local minimizer of this functional satisfies

$$\Delta u = 0 \quad \text{in } \{u > 0\}, \qquad |\nabla u| = Q \quad \text{on } \partial\{u > 0\}.$$

A volume-constrained version of this problem was analyzed in [1] and [42, 88].

2.2.2 Time-Dependent Problem

To introduce the time-dependent problem, we adopt Hamilton's principle. We thus first define a Lagrangian and the corresponding action for our phenomenon. The action integral can be expressed as

$$I_{\text{adh}}(u) = \int_0^T \int_{\Omega} \left((u_t)^2 \chi_{\{u>0\}} - |\nabla u|^2 - Q^2 \chi_{\{u>0\}} \right) dx \, dt, \tag{2.8}$$

where T is a positive constant, and we have omitted the coefficient $S/2$ in (2.7). We remark that the first term is the kinetic energy of the tape in the vertical direction.

To find a stationary point of the action integral, we need to derive the momentum equation. However, the discontinuity of the characteristic function appearing in the functional prevents us from calculating the first variation in the usual manner. To bypass this complication, to begin with, we assume that the stationary point u and its free boundary $\partial\{u > 0\}$ are sufficiently regular. Then a stationary point u should be nonnegative. This can be rigorously justified using the maximum principle for the Laplacian. Indeed, if u is continuous and there is a point (x, t) such that $u(x, t) < 0$, we observe that $\Delta u = 0$ in $\{u(\cdot, t) < 0\}$, which by maximum principle implies that $u(x, t)$ must be equal to zero, a contradiction. Therefore, let us consider a nonnegative stationary point u of I_{adh} in $C^0(Q_T) \cap W^{1,2}(Q_T)$, where Q_T is as in (2.1). Then for any $\zeta \in C_0^\infty(Q_T \cap \{u > 0\})$ we get

$$\int_{Q_T \cap \{u > 0\}} (u_t \zeta_t - \nabla u \cdot \nabla \zeta) \, dx \, dt = 0. \tag{2.9}$$

This means that any stationary point u satisfies the wave equation in $\{u > 0\}$ and equals zero otherwise.

On the other hand, to obtain the free boundary condition, we calculate the so-called *inner variation*. To this end, let us assume that $u \in C^2(Q_T \cap \{u > 0\})$ and that the free boundary $\partial\{u > 0\}$ is a C^1-surface. For arbitrary $\eta \in C_0^\infty(Q_T; \mathbb{R}^m \times \mathbb{R})$ and $\varepsilon > 0$ satisfying $0 < \varepsilon < \text{dist}(\text{spt} \, \eta, \partial Q_T)$, we consider the map $\tau_\varepsilon : Q_T \to Q_T$ defined by $\tau_\varepsilon(z) := z + \varepsilon\eta(z)$. Here z denotes the space-time variable, i.e., $z_j = x_j$ ($j = 1, \ldots, m$) and $z_{m+1} = t$, and spt η designates the support of η. Now we set $u_\varepsilon(z) = u \circ \tau_\varepsilon^{-1}(z)$ and formally calculate the inner variation $(d/d\varepsilon)I_{adh}(u_\varepsilon)|_{\varepsilon=0} = 0$.

By change of variables,

$$I_{adh}(u_\varepsilon) = -\int_{Q_T \cap \{u > 0\}} \left(\left((D\tau_\varepsilon)^{-1}\nabla_z^+ u\right) \cdot \left((D\tau_\varepsilon)^{-1}\nabla_z^- u\right) + Q^2 \right) |\det D\tau_\varepsilon| \, dz,$$

where $\nabla_z^+ f = (f_{z_1}, \cdots, f_{z_m}, f_{z_{m+1}})$, $\nabla_z^- f = (f_{z_1}, \cdots, f_{z_m}, -f_{z_{m+1}})$, "$\cdot$" is the inner product in $\mathbb{R}^{m+1}$, and $D\tau_\varepsilon$ is the Jacobi matrix of the map τ_ε.

It follows that

$$I_{adh}(u_\varepsilon) - I_{adh}(u) = -\varepsilon \int_{Q_T \cap \{u > 0\}} \left(\nabla_z^+ u \cdot \nabla_z^- u + Q^2 \right) (\nabla_z^+ \cdot \eta) \, dz$$

$$+ \varepsilon \int_{Q_T \cap \{u > 0\}} 2\nabla_z^+ u \, (D\eta) \, \nabla_z^- u \, dz + o(\varepsilon).$$

Hence, using the identity

$$\nabla_z^+ \cdot \left(\eta \left(\nabla_z^+ u \cdot \nabla_z^- u\right) - 2 \left(\eta \cdot \nabla_z^+ u\right) \nabla_z^- u \right)$$

$$= (\nabla_z^+ \cdot \eta) \left(\nabla_z^+ u \cdot \nabla_z^- u\right) - 2\nabla_z^+ u \, (D\eta) \, \nabla_z^- u - 2 \left(\eta \cdot \nabla_z^+ u\right) (\Delta u - u_{tt}),$$

and the fact following from (2.9) that $u_{tt} - \Delta u = 0$ in $\{u > 0\}$, we have

$$\frac{d}{d\varepsilon} I_{adh}(u_\varepsilon)|_{\varepsilon=0}$$

$$= -\int_{Q_T \cap \{u > 0\}} \nabla_z^+ \cdot \left(\eta \left(\nabla_z^+ u \cdot \nabla_z^- u\right) - 2 \left(\eta \cdot \nabla_z^+ u\right) \nabla_z^- u + Q^2 \eta \right) \, dz$$

$$= -\int_{Q_T \cap \partial\{u > 0\}} \left(\left((\nabla_z^+ u \cdot \nabla_z^- u) \eta - 2 \left(\eta \cdot \nabla_z^+ u\right) \nabla_z^- u\right) \cdot \nu + Q^2 (\eta \cdot \nu) \right) \, d\mathcal{H}^m,$$

where $\mathscr{H}^m$ is the m-dimensional Hausdorff measure, and ν is the outer unit normal vector to $Q_T \cap \{u > 0\}$. On the other hand, since $\nabla_z^+ u = -\nu|\nabla_z^+ u|$, we get

$$\frac{d}{d\varepsilon} I_{\text{adh}}(u_\varepsilon)|_{\varepsilon=0} = \int_{Q_T \cap \partial\{u>0\}} \left((\nabla_z^+ u \cdot \nabla_z^- u) - Q^2 \right) (\eta \cdot \nu) \, d\mathscr{H}^m.$$

This is equal to 0 for any $\eta \in C_0^\infty(Q_T; \mathbb{R}^m \times \mathbb{R})$, leading to the free boundary condition $|\nabla u|^2 - (u_t)^2 = Q^2$ on $\partial\{u > 0\}$, which is a generalization of (2.4). Rewriting the equations in the strong form, we have

$$\begin{cases} u_{tt} - \Delta u & = 0 \quad \text{in} \quad Q_T \cap \{u > 0\}, \\ |\nabla u|^2 - (u_t)^2 & = Q^2 \quad \text{on} \quad Q_T \cap \partial\{u > 0\}. \end{cases} \tag{2.10}$$

For this problem, we have shown the existence of a strong solution in the one-dimensional case. The following theorem outlines this result—see [35] for details.

Theorem 2.1 (K. Kikuchi and S. Omata [35]) *Let $\Omega = (0, \infty)$. Suppose that a boundary condition $u(0, t) = g(t)$ with $g'(t) > 0$ is given, and assume that the initial datum $u_0(x) = u(x, 0)$ is positive on some nonempty interval $(0, L)$. Moreover, assume that at $x = 0$ and $x = L$, the initial data satisfy appropriate compatibility conditions. Then the initial-boundary value problem for (2.10) has a unique time-local classical solution $u \in C^2(Q_T \cap \{u > 0\})$ and the free boundary $\partial\{u > 0\}$ is a C^1-curve.*

Attention to the work of G. Lazzaroni et al. [18, 39–41] should also be noted. The authors have obtained existence and uniqueness results in the one-dimensional setting for a related model which describes the dynamic debonding of thin films. In their model, the local strength of the adhesive between the film and the substrate depends on the speed of the debonding front. A wave equation, holding in the debonded region, is coupled with a dynamic free boundary condition (Griffith's criterion) for describing the evolution of the front. Explicit solutions to a number of specific problems were also obtained, and the authors analyzed the zero-inertia limit of solutions which illustrated interesting properties of the propagation. In relation to vibrations with adhesion, we would like to mention also the work of [15, 16], where it is modeled by a semilinear wave equation with a discontinuous source term, and to [17], where the analysis is extended to fourth-order differential operators.

2.2.3 Regularized Equation and Its Limit

In this section, we present a regularization approach to the problem of non-differentiable characteristic function $\chi_{\{u>0\}}$ present in the action functional. This approach has been employed in the elliptic case, where a smoothing of the characteristic function allows for the standard derivation of the Euler–Lagrange equation and recovery of the free boundary condition in the limit of smoothing parameter [10]. We follow this method in the setting of hyperbolic equations.

For $\varepsilon > 0$, let β_ε be a smooth function satisfying

$$\beta_\varepsilon(s) \begin{cases} = 0 & s \leq 0, \\ \leq 1/\varepsilon & 0 < s < \varepsilon \\ = 0 & \varepsilon \leq s, \end{cases} \quad \text{with} \quad |\beta'_\varepsilon(s)| \leq \frac{C}{\varepsilon^2} \quad \text{for some } C > 0,$$

and $\int_0^\varepsilon \beta_\varepsilon(s)\,ds = \frac{1}{2}$. We then have

$$B_\varepsilon(u) := \int_0^u \beta_\varepsilon(s)\,ds \xrightarrow[\varepsilon \to 0+]{} \begin{cases} \frac{1}{2} & \text{if } u > 0 \\ 0 & \text{if } u \leq 0 \end{cases}$$

pointwise, hence we can think of $2B_\varepsilon(u)$ as an approximation of $\chi_{\{u>0\}}$.

Now we introduce the smoothed action functional. Let Γ be a positive constant which controls the contact angle. We modify the action integral (2.8) to

$$I_\varepsilon(u) = \int_0^T \int_\Omega \left((u_t)^2 \chi_{\{u>0\}} - |\nabla u|^2 - 2\Gamma B_\varepsilon(u) \right) dx\,dt, \tag{2.11}$$

and calculate the necessary condition for its stationarity. Written in the strong form, this condition reads

$$\chi_{\overline{\{u>0\}}} u_{tt} = \Delta u - \Gamma \beta_\varepsilon(u) \quad \text{in } Q_T. \tag{2.12}$$

We will show through formal calculations that this equation is a correct approximation of the original singular problem.

Proposition 2.1 *Let β_ε and B_ε be the functions defined above. Suppose that for every $\varepsilon > 0$ there is u^ε, a solution to (2.12). Moreover, we assume that u^ε converges to some v as $\varepsilon \to 0+$ in an appropriate topology, where v satisfies $\Delta v - v_{tt} = 0$ in $Q_T \cap \{v > 0\}$. Then $|\nabla v|^2 - (v_t)^2 = \Gamma$ on $\partial\{v > 0\}$.*

Proof We multiply both sides of Eq. (2.12) by ζu_k^ε, where u_k^ε denotes the derivative $\partial u^\varepsilon / \partial z_k$ ($k = 1, \ldots, m+1$), and $\zeta \in C_0^\infty(Q_T)$. Next, we integrate over Q_T to get

$$\int_{Q_T} \zeta u_k^\varepsilon (\Delta u^\varepsilon - \chi_{\{u^\varepsilon>0\}} u_{tt}^\varepsilon)\,dz = \int_{Q_T} \Gamma \zeta u_k^\varepsilon \beta_\varepsilon(u^\varepsilon)\,dz. \tag{2.13}$$

Noting that $\left[B_\varepsilon(u)\right]_{x_k} = \beta_\varepsilon(u)u_k$, assuming that $2B_\varepsilon(u^\varepsilon) \to \chi_{\{v>0\}}$ and using integration by parts, the right-hand side of (2.13) can be calculated as

$$
\int_{Q_T} \Gamma \zeta u_k^\varepsilon \beta_\varepsilon(u^\varepsilon)\, dz = -\int_{Q_T} \Gamma \zeta_k B_\varepsilon(u^\varepsilon)\, dz
$$

$$
\xrightarrow[\varepsilon\to 0+]{} -\frac{1}{2}\int_{Q_T\cap\{v>0\}} \Gamma \zeta_k\, dz = -\frac{1}{2}\int_{Q_T\cap\partial\{v>0\}} \Gamma \zeta\, v_k\, d\mathscr{H}^m.
$$

Here $v = (v_1 \cdots v_{m+1})$ is the unit outer normal to $\{v>0\}$ within Q_T and $\mathscr{H}^m$ is a shorthand for the restriction $\mathscr{H}^m\llcorner\,\partial\{v>0\}$ of $\mathscr{H}^m$ to the free boundary $\partial\{v>0\}$. On the other hand, the left-hand side of (2.13) becomes

$$
\int_{Q_T} \zeta u_k^\varepsilon\left(\Delta u^\varepsilon - \chi_{\{u^\varepsilon>0\}}u_{tt}^\varepsilon\right) dz = -\int_{Q_T}\left(\nabla(\zeta u_k^\varepsilon)\cdot\nabla u^\varepsilon - (\zeta u_k^\varepsilon)_t u_t^\varepsilon \chi_{\{u^\varepsilon>0\}}\right) dz.
$$

We have assumed that u^ε touches the free boundary smoothly enough and thus $(\zeta u_t^\varepsilon \chi_{\{u^\varepsilon>0\}})_t = (\zeta u_t^\varepsilon)_t \chi_{\{u^\varepsilon>0\}}$. Moreover, outside of $\{u^\varepsilon>0\}$ the function u^ε is zero. The calculation of the left-hand side of (2.13) then continues as

$$
= -\int_{Q_T\cap\{u^\varepsilon>0\}}\left(\left[\nabla u^\varepsilon\cdot\nabla\zeta - u_t^\varepsilon\zeta_t\right]u_k^\varepsilon + \frac{1}{2}\frac{\partial}{\partial z_k}\left[|\nabla u^\varepsilon|^2 - (u_t^\varepsilon)^2\right]\zeta\right) dz
$$

$$
= -\int_{Q_T\cap\{u^\varepsilon>0\}}\left(\left[\nabla u^\varepsilon\cdot\nabla\zeta - u_t^\varepsilon\zeta_t\right]u_k^\varepsilon - \frac{1}{2}\left[|\nabla u^\varepsilon|^2 - (u_t^\varepsilon)^2\right]\zeta_k\right) dz
$$

$$
\xrightarrow[\varepsilon\to 0+]{} -\int_{Q_T\cap\{v>0\}}\left([\nabla v\cdot\nabla\zeta - v_t\zeta_t]v_k - \frac{1}{2}\left[|\nabla v|^2 - (v_t)^2\right]\zeta_k\right) dz
$$

$$
= -\int_{Q_T\cap\{v>0\}}\left(\nabla(\zeta v_k)\cdot\nabla v - (\zeta v_k)_t v_t\right) dz + \frac{1}{2}\int_{Q_T\cap\partial\{v>0\}}\left(|\nabla v|^2 - (v_t)^2\right)\zeta v_k\, d\mathscr{H}^m
$$

$$
= \int_{Q_T\cap\{v>0\}} \zeta v_k(\Delta v - v_{tt})\, dz - \int_{Q_T\cap\partial\{v>0\}} \zeta v_k\,(\nabla v, -v_t)\cdot v\, d\mathscr{H}^m
$$

$$
+ \frac{1}{2}\int_{Q_T\cap\partial\{v>0\}}\left(|\nabla v|^2 - (v_t)^2\right)\zeta v_k\, d\mathscr{H}^m
$$

$$
= -\frac{1}{2}\int_{Q_T\cap\partial\{v>0\}}\left[|\nabla v|^2 - (v_t)^2\right]\zeta v_k\, d\mathscr{H}^m.
$$

The last equality follows from the fact that the unit outer normal to $\{v>0\}$ can be written as $v = -Dv/|Dv|$, so that $v_k = -v_k|Dv|$ on $Q_T \cap \partial\{v>0\}$. Here, $Dv = (\frac{\partial v}{\partial x_1}, \cdots, \frac{\partial v}{\partial x_m}, \frac{\partial v}{\partial t})$. Thus we arrive at

$$
\frac{1}{2}\int_{Q_T\cap\partial\{v>0\}}\left[|\nabla v|^2 - (v_t)^2\right]\zeta v_k\, d\mathscr{H}^m = \frac{1}{2}\int_{Q_T\cap\partial\{v>0\}} \Gamma \zeta v_k\, d\mathscr{H}^m,
$$

which, by the arbitrariness of ζ, leads to

$$|\nabla v|^2 - (v_t)^2 = \Gamma \quad \text{on } \partial\{v > 0\}.$$

$\square$

We conclude that if we set $\Gamma = Q^2$, Eq. (2.12) is a reasonable approximation of the tape peeling problem.

Furthermore, it is of interest to investigate the limit $\varepsilon \to 0+$ in the PDE (2.12) itself. Here we assume that the angle $|Dv|$ is nonzero on $\partial\{v > 0\}$. If u^ε converges to v in an appropriate sense, and moreover $\mathcal{H}^m(\partial\{v > 0\})$ is finite, by testing Eq. (2.12) by ζ as above, taking the limit as $\varepsilon \to 0+$ and applying Green's theorem, we have

$$\lim_{\varepsilon \to 0+} \int_{Q_T} \Gamma \beta_\varepsilon(u^\varepsilon) \zeta \, dz = \lim_{\varepsilon \to 0+} \int_{Q_T} (-\nabla u^\varepsilon \cdot \nabla \zeta + u_t^\varepsilon \zeta_t) \, dz$$

$$= \int_{Q_T \cap \{v > 0\}} (-\nabla v \cdot \nabla \zeta + v_t \zeta_t) \, dz$$

$$= \int_{Q_T \cap \{v > 0\}} (\Delta v - v_{tt}) \zeta \, dz + \int_{\partial\{v > 0\}} (-\nabla v, v_t) \cdot v \, \zeta \, d\mathcal{H}^m$$

$$= \int_{\partial\{v > 0\}} \frac{|\nabla v|^2 - v_t^2}{|Dv|} \zeta \, d\mathcal{H}^m$$

$$= \int_{\partial\{v > 0\}} \frac{\Gamma}{|Dv|} \zeta \, d\mathcal{H}^m,$$

where the last equality follows from Proposition 2.1. We deduce that formally

$$\beta_\varepsilon(u^\varepsilon) \to \frac{1}{|Dv|} \mathcal{H}^m \llcorner \partial\{v > 0\} \quad \text{as } \varepsilon \to 0+.$$

The above arguments yield the following degenerate hyperbolic equation:

$$\chi_{\overline{\{u > 0\}}} u_{tt} - \Delta u = -\frac{Q^2}{|Du|} \mathcal{H}^m \llcorner \partial\{u > 0\}. \tag{2.14}$$

We observe that when $u > 0$, a solution u to (2.14) satisfies the wave equation, while when $u < 0$, the equation reduces to $\Delta u = 0$ for a.e. $t \in (0, T)$, hence by the maximum principle u is identically zero in $\{u \leq 0\}$.

In order to confirm the compatibility of the degenerate Eq. (2.14) with the formulation (2.10), it remains to check the free boundary condition. Writing Eq. (2.14) in the weak form, we have

$$\int_0^T \int_\Omega \left(\left(\chi_{\overline{\{u > 0\}}} \zeta \right)_t u_t - \nabla u \cdot \nabla \zeta \right) dx \, dt = \int_{\partial\{u > 0\}} \frac{Q^2}{|Du|} \zeta \, d\mathcal{H}^m, \quad \zeta \in C_0^\infty(Q_T).$$

We split the integral on the left-hand side into four parts, namely, into integrals over the regions $Q_T \cap \{u > 0\}$, $Q_T \cap \partial\{u > 0\}$, $Q_T \cap \partial\{u = 0\}^\circ$ and $Q_T \cap \{u = 0\}^\circ$. Integrating by parts, we find that all these integrals except for the one over $Q_T \cap \partial\{u > 0\}$ vanish. Since $\chi_{\overline{\{u>0\}}} = 1$ on $\partial\{u > 0\}$, the left-hand side thus equals to

$$\int_{\partial\{u>0\}} \left(\chi_{\overline{\{u>0\}}} \zeta u_t \frac{-u_t}{|Du|} + \nabla u \cdot \frac{\nabla u}{|Du|} \zeta \right) d\mathcal{H}^m = \int_{\partial\{u>0\}} \frac{|\nabla u|^2 - u_t^2}{|Du|} \zeta \, d\mathcal{H}^m,$$

and we have successfully, though formally, recovered the free boundary condition $|\nabla u|^2 - u_t^2 = Q^2$ on $\partial\{u > 0\}$. The above analysis is shown to work for the elliptic [10] and parabolic [14] problems, but a rigorous proof for the hyperbolic setting is still wanting.

2.3 Droplet Motion: A Volume-Preserving Problem

In this section, we consider problems with volume constraint. The motivation is to analyze the motion of a membrane that encloses a constant volume [84, 85, 90].

2.3.1 Volume-Constrained Problem Without Free Boundary

Before treating the corresponding free boundary problem, we first analyze the vibration of an elastic membrane covering a container filled with an incompressible liquid. We allow an outer force F to act on the membrane, but assume the existence of its potential P. We describe the shape of the film by the graph of a scalar function u defined on a domain Ω.

To obtain the equation of motion, we employ the action integral of the system. Let ρ be the surface density of the film and γ its elastic coefficient. Both are assumed to be constant. Here we ignore the effect of any fluid flow inside the container, leaving a more detailed discussion of this topic to Sect. 2.3.3. We can write the Lagrangian for the system as

$$L_{\text{vol}}(u) = \int_\Omega \left(\frac{1}{2}\rho(u_t)^2 - \frac{1}{2}\gamma|\nabla u|^2 + P(u) \right) dx. \tag{2.15}$$

Then the action integral on a time interval $(0, T)$ becomes $I_{\text{vol}}(u) = \int_0^T L_{\text{vol}}(u) \, dt$, and the equation of motion is the associated Euler–Lagrange equation.

For some $V > 0$, given the volume constraint condition

$$\int_\Omega u(x, t) \, dx = V \qquad \forall t \in [0, T], \tag{2.16}$$

it is natural to define the set of admissible functions

$$\mathcal{K}_{\text{vol}}^T := \left\{ u \in H^1(Q_T); \ u(x,0) = u_0(x), \ u_t(x,0) = v_0(x), \ u|_{\partial\Omega} = 0, \ \int_\Omega u \, dx = V \right\},$$

inside which we will seek for a stationary point. Here, u_0 is the initial shape of the membrane, v_0 is its initial velocity, and, as before, $Q_T := \Omega \times (0, T)$. Note that, for the sake of simplicity, we have prescribed homogeneous boundary conditions.

Let us calculate the first variation of $I_{\text{vol}}(u)$. Test functions should be chosen so that the perturbations u_ε belong to $\mathcal{K}_{\text{vol}}^T$. Since the volume of a test function $\varphi \in C_0^\infty(Q_T)$ is given by $\Phi(t) = \int_\Omega \varphi(x,t) \, dx$, and therefore, $\int_\Omega (u + \varepsilon\varphi) \, dx = V + \varepsilon\Phi$, it is natural to take perturbations of the form

$$u_\varepsilon = V \frac{u + \varepsilon\varphi}{V + \varepsilon\Phi} = \frac{u + \varepsilon\varphi}{1 + \frac{\varepsilon}{V}\Phi}.$$

Then the volume is preserved and, for small ε, the denominator stays positive.

Writing out the necessary condition for the stationary point $(d/d\varepsilon)I_{\text{vol}}(u_\varepsilon)|_{\varepsilon=0} = 0$, we have

$$0 = \lim_{\varepsilon \to 0} \frac{1}{\varepsilon} \Bigg[\int_{Q_T} \frac{\rho}{2} \bigg(\Big(\frac{(u_t + \varepsilon\varphi_t)(1 + \frac{\varepsilon}{V}\Phi) - (u + \varepsilon\varphi)\frac{\varepsilon}{V}\Phi_t}{(1 + \frac{\varepsilon}{V}\Phi)^2} \Big)^2 - (u_t)^2 \bigg) \, dx \, dt$$

$$- \int_{Q_T} \frac{\gamma}{2} \Big(\frac{|\nabla u + \varepsilon\nabla\varphi|^2}{(1 + \frac{\varepsilon}{V}\Phi)^2} - |\nabla u|^2 \Big) \, dx \, dt + \int_{Q_T} \Big(P\Big(\frac{u + \varepsilon\varphi}{1 + \frac{\varepsilon}{V}\Phi} \Big) - P(u) \Big) \, dx \, dt \Bigg]$$

$$= \int_{Q_T} \Big[\rho\Big(u_t\varphi_t - \frac{1}{V}u_t(\Phi u)_t \Big) - \gamma\Big(\nabla u \cdot \nabla\varphi - \frac{\Phi}{V}|\nabla u|^2 \Big) + P'(u)\Big(\varphi - \frac{\Phi}{V}u \Big) \Big] \, dx \, dt$$

$$= \int_{Q_T} \big[\rho u_t\varphi_t - \gamma\nabla u \cdot \nabla\varphi + P'(u)\varphi \big] \, dx \, dt$$

$$+ \frac{1}{V} \int_{Q_T} \big[-\rho u_t(u\Phi)_t + \gamma|\nabla u|^2\Phi - P'(u)u\Phi \big] \, dx \, dt.$$

If u is smooth enough, then the last integral can be rearranged by using integration by parts:

$$\frac{1}{V} \int_{Q_T} \Big(\rho u_{tt}u + \gamma|\nabla u|^2 + F(u)u \Big) \Phi \, dx \, dt,$$

where we have used the definition of the potential $F(u) = -P'(u)$. Hence, defining the time-dependent function λ_u by

$$\lambda_u(t) = \frac{1}{V} \int_\Omega \Big(\rho u_{tt}u + \gamma|\nabla u|^2 + F(u)u \Big) \, dx, \tag{2.17}$$

we get the equation

$$\int_{Q_T} (-\rho u_t \varphi_t + \gamma \nabla u \cdot \nabla \varphi + F(u)\varphi - \lambda_u(t)\varphi) \, dx \, dt = 0 \qquad \forall \varphi \in C_0^\infty(Q_T).$$

The pointwise form is

$$\rho u_{tt} = \gamma \Delta u - F(u) + \lambda_u(t), \tag{2.18}$$

which is simply the wave equation with a nonlocal term representing a Lagrange multiplier. This equation can be also obtained through application of the Lagrange multiplier method, i.e., we seek a stationary point of the action integral

$$\widetilde{I}_{\text{vol}}(u) = \int_0^T L_{\text{vol}}(u) \, dt + \int_0^T \lambda_u(t)\left(\int_\Omega u \, dx\right) dt,$$

in the admissible set $\mathcal{K}^T$ which is obtained by deleting the constraint from the definition of the set $\mathcal{K}_{\text{vol}}^T$.

For this problem we have proved the following well-posedness result.

Theorem 2.2 (K. Svadlenka and S. Omata [84]) *Suppose that the initial data $u_0, v_0 \in H^2(\Omega)$ satisfy compatibility conditions $u_0(x) = v_0(x) = 0$ for $x \in \partial\Omega$ and $\int_\Omega u_0 \, dx = V$, $\int_\Omega v_0 \, dx = 0$. Further suppose that $F \equiv 0$ or that F satisfies certain growth conditions detailed in [84]. Then the initial-boundary value problem for (2.18) with λ_u given in (2.17) has unique solution $u \in W^{2,\infty}(0, T; L^2(\Omega)) \cap L^\infty(0, T; H_0^1(\Omega))$ satisfying the volume constraint (2.16).*

2.3.2 Volume-Constrained Problem with Free Boundary

We extend the problem from the previous section to the setting including a free boundary. The resulting model can be then used to describe phenomena such as droplet motion on a plane, or bubble motion on a water surface. In our analysis, we restrict the surface to be given by the graph of a scalar function [90]. The strategy is the same as in the previous section, that is, we compute a stationary point of the action integral and obtain an equation of motion. Since there is a volume constraint, a Lagrange multiplier appears and acts as a uniform outer force on the area where the membrane is detached from the obstacle.

The free boundary condition for a stationary droplet is given by the well-known Young's law. In particular, denoting the surface tensions of the three types of interfaces as γ_g (liquid–gas), γ_{sl} (solid–liquid) and γ_{sg} (solid–gas), see also Fig. 2.3, the balance of forces $\gamma_g \cos\theta = \gamma_{sg} - \gamma_{sl}$ holds, yielding the equilibrium contact angle θ [20]. In the case of dynamic contact angles, computation of the inner variation of the action with the force term $Q^2 \chi_{\{u>0\}}$ gives a free boundary condition

Fig. 2.3 Setup and notation
for the moving droplet
problem

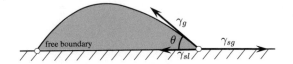

of the same form (2.10) as in the tape peeling problem. This implies that the volume
constraint has no impact on the free boundary condition.

When we smooth the term $Q^2 \chi_{\{u>0\}}$ by $2\Gamma B_\varepsilon(u)$, where $\Gamma = Q^2$, and
subsequently compute the first variation of the action functional, we arrive at the
equation

$$\chi_{\overline{\{u>0\}}} u_{tt} = \Delta u - \Gamma \beta_\varepsilon(u) + \lambda(u) \chi_{\{u>0\}}, \quad \lambda(u) = \frac{1}{V} \int_\Omega \left(u_{tt} u + |\nabla u|^2 + \Gamma \beta_\varepsilon(u) u \right) dx.$$
(2.19)

We have scaled out physical coefficients in order to make the formulas more concise.
We again note that solutions to this equation possess a free boundary $\partial\{u > 0\}$ due
to the characteristic function in the Lagrange multiplier term.

A weak solution to (2.19) is defined as follows:

Definition 2.1 $u \in H^1(0, T; L^2(\Omega)) \cap L^\infty(0, T; H_0^1(\Omega))$ is a weak solution to
Eq. (2.19) with initial conditions $u(x, 0) = u_0(x)$, $u_t(x, 0) = v_0(x)$ and boundary
condition $u = 0$ on $\partial\Omega$, if $u = 0$ outside of $\{u > 0\}$, and if for all $\phi \in C_0^\infty((\Omega \times [0, T)) \cap \{u > 0\})$ and for all $\tilde{u} \in C_0^\infty((\Omega \times [0, T]) \cap \{u > 0\})$ with $\int_\Omega \tilde{u}(x, t) \, dx = V$,

$$\int_0^T \int_\Omega (-u_t \phi_t + \nabla u \cdot \nabla \phi + \Gamma \beta_\varepsilon(u)\phi) \, dx \, dt - \int_\Omega v_0 \phi(0, x) \, dx$$

$$= \frac{1}{V} \int_0^T \int_\Omega (-u_t(\tilde{u}\Phi)_t + (\nabla u \cdot \nabla \tilde{u} + \Gamma \beta_\varepsilon(u)\tilde{u})\Phi) \, dx \, dt - \frac{1}{V} \int_\Omega v_0 \tilde{u}(x, 0) \Phi(0) \, dx$$

is satisfied. Here, $\Phi(t) = \int_\Omega \phi(x, t) \, dx$.

This problem has been solved in one spatial dimension, as stated in the following
theorem.

Theorem 2.3 (E. Ginder and K. Svadlenka [26]) *Let Ω be a one-dimensional
domain and let u_0 and v_0 belong to $H_0^1(\Omega)$. Then there exists a weak solution
to (2.19) in the sense of* Definition 2.1.

In Chap. 4 we provide a more detailed discussion, including the higher-
dimensional case. We also remark that the model described above has been applied
to investigating the motion of droplets on a surface treated with a surfactant, see,
e.g., [57].

2.3.3 A Coupled Model of Membrane and Fluid Motion

In applications, accounting for the interaction of the membrane with the enclosed fluid is often required when precisely expressing the phenomenon of a moving membrane or droplet. Here we present a model of membrane motion coupled with an incompressible fluid that is enveloped by the membrane. We again assume, for the sake of simplicity, that the membrane can be described by the graph of a scalar function u, and moreover, we consider the two-dimensional setting. For a given $L > 0$, the membrane is defined on the spatial interval $x \in [0, L]$ and the fluid then occupies at time t the region $\Omega(t)$ between the membrane and a flat obstacle positioned at $u = 0$, i.e.,

$$\Omega(t) = \{(x, y) \mid x \in (0, L), \ y \in (0, u(x, t))\}.$$

Under these assumptions, we develop a mathematical and numerical model through weak coupling. In a weakly coupled problem, the equation of motion for the membrane and those for the fluid are solved alternately, each for a short time, and the result of one equation is then used as the input for the other.

The equation for the membrane was derived in the beginning of this section and reads

$$\sigma u_{tt} = \gamma u_{xx} + f_p + f_v + f + \lambda, \tag{2.20}$$

$$\lambda = \frac{1}{V} \int_0^L \left(\sigma u u_{tt} + \gamma |u_x|^2 - (f_p + f_v + f)u \right) dx. \tag{2.21}$$

Here σ, γ, V, f stand for surface density of the membrane, tension coefficient of the membrane, total volume of the fluid and an arbitrary outer force, respectively. The forces f_p, f_v are due to pressure and viscous stress of the fluid, and have the form

$$f_p(x, t) = p(x, u(x, t), t),$$

$$f_v(x, t) = \mu \left[u_x \left(\frac{\partial(v \cdot \hat{y})}{\partial x} + \frac{\partial(v \cdot \hat{x})}{\partial y} \right) - 2 \frac{\partial(v \cdot \hat{y})}{\partial y} \right] \Big|_{y=u(x,t)},$$

where $p(x, y, t)$ is the pressure, $v(x, y, t)$ is the velocity at a point (x, y), μ is the viscosity of the fluid, and the unit vectors $\hat{x}, \hat{y}$ are defined by $\hat{x} = (1, 0), \hat{y} = (0, 1)$.

To model the fluid flow, we solve the incompressible Navier–Stokes system

$$\rho(v_t + (v \cdot \nabla)v) = -\nabla p + \mu \Delta v \qquad \text{in } \Omega(t),$$

$$\nabla \cdot v = 0 \qquad \text{in } \Omega(t),$$

where ρ denotes fluid density, equipped with suitable initial conditions and the following no-slip and free-slip boundary conditions:

$$\boldsymbol{v} = \boldsymbol{0} \qquad \text{at} \quad y = 0$$

$$\boldsymbol{v} \cdot \boldsymbol{n} = \frac{u_t}{\sqrt{1 + u_x^2}}, \qquad \frac{\partial (\boldsymbol{v} \cdot \boldsymbol{\tau})}{\partial \boldsymbol{n}} = 0 \qquad \text{at} \quad y = u(x, t).$$

Here, $\boldsymbol{n}$ denotes the unit outer normal to $\Omega(t)$ and $\boldsymbol{\tau}$ is its unit tangent vector.

For illustration, let us show results of a couple of numerical experiments. In the first example, we consider the fluid flow caused by pushing on the membrane which encloses it (see [31] for details). In order to solve the hyperbolic equation (2.20) with a constraint, we used the discrete Morse flow (explained in the following chapter), while the coupling with the Navier–Stokes equations was resolved by the MPS (Moving Particle Semi-implicit) method [37]. Figure 2.4 depicts the interaction between a vibrating membrane and an enclosed fluid, which eventually induces the flow of the fluid towards the right. In the simulation, an outer force has been applied to the membrane at the center of the domain.

Another interesting application of this method is to the computation of droplet motion with free boundary [70]. Figure 2.5 shows a water droplet sliding on the bottom of an inclined plane. Due to surface tension forces, the droplet does not detach from the plane but, if the gravity is strong enough, slides down the plane. To implement the simulation, we use the discrete Morse flow to solve the hyperbolic free boundary problem introduced in Sect. 2.3.2, and the SPH (Smoothed Particle Hydrodynamics) method [55] to solve the Euler equations for the fluid in three dimensions.

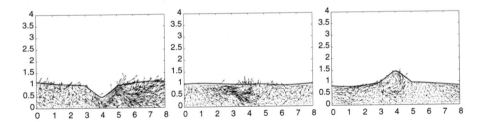

Fig. 2.4 Simulation of fluid flow driven by a vibrating membrane

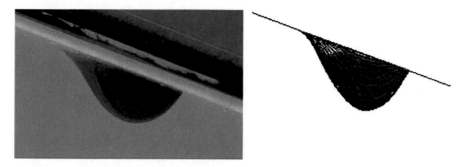

Fig. 2.5 (Left) A real photograph of a water drop sliding down over the bottom of an inclined surface. (Right) A snapshot of the numerical result

Chapter 3
Discrete Morse Flow

Having given some motivation for studying free boundary problems of hyperbolic type, we proceed to introducing a promising mathematical tool for their analysis and numerical approximation. The approach is often called *hyperbolic discrete Morse flow* and is a hyperbolic analogy of Rothe's method [77] which is well established in the field of parabolic equations.

3.1 Discrete Morse Flow for the Linear Wave Equation

We shall first explain the idea behind the discrete Morse flow on the simple example of the wave equation. We consider a bounded domain $\Omega \subset \mathbb{R}^m$ with smooth boundary $\partial\Omega$, on which the homogeneous Dirichlet boundary condition is given, and a time interval $(0, T)$, $T > 0$. For an initial function $u_0 \in H_0^1(\Omega)$ and an initial velocity $v_0 \in H_0^1(\Omega)$, we have the following initial-boundary value problem:

$$u_{tt}(x, t) = \Delta u(x, t) \quad \text{in } Q_T := \Omega \times (0, T), \tag{3.1}$$

$$u(x, t) = 0 \quad \text{on } \partial\Omega \times (0, T), \tag{3.2}$$

$$u(x, 0) = u_0(x) \quad \text{in } \Omega, \tag{3.3}$$

$$u_t(x, 0) = v_0(x) \quad \text{in } \Omega. \tag{3.4}$$

To investigate the well-posedness of this problem, several techniques have been developed. Here we explain a relatively recent method that has substantial potential for generalization to more complex problems and for application to numerical approximation. The method is based on a semi-discretization in time. First, we fix a natural number $N > 0$ that determines the time step $h = T/N$, and set $u_{-1}(x) = u_0(x) - hv_0(x)$. The function u_0 corresponds to the approximate solution at time level $t = 0$, while function u_{-1} is the approximate solution at time level $t = -h$. We define an approximate solution u_n on subsequent time levels $t = nh$ for $n = 1, 2, \ldots, N$, to be a minimizer of the following functional in $H_0^1(\Omega)$:

$$J_n(u) = \int_\Omega \frac{|u - 2u_{n-1} + u_{n-2}|^2}{2h^2}\, dx + \frac{1}{2}\int_\Omega |\nabla u|^2\, dx. \tag{3.5}$$

We observe that the second term of the functional is lower-semicontinuous with respect to the weak topology in $H^1(\Omega)$ and the first term is continuous in $L^2(\Omega)$. The existence of a unique minimizer in $H_0^1(\Omega)$ then follows immediately by the direct method from the fact that the functionals are convex in u, ∇u and bounded below for each $n = 1, 2, \ldots, N$. This is a crucial advantage over the continuous version of this functional, the Lagrangian. Of course, if other terms, representing outer forces, etc., are present, we have to make certain assumptions concerning these terms in order to prove the existence minimizers.

Here we pause to mention that another interesting approach to recast a nonlinear wave-type evolution as a minimization problem was proposed by E. De Giorgi and analyzed in [81]. This method does not discretize time but instead employs a singular weight in the integrand, which again conveniently leads to a convex minimization problem. The idea was further exploited in [19, 82].

Next, we define approximate solutions $\bar{u}^h$ and u^h through interpolation of the minimizers $\{u_n\}_{n=-1}^N$ in time. The interpolation is given by

$$\begin{aligned}
\bar{u}^h(x, t) &= u_n(x), \\
u^h(x, t) &= \frac{t - (n-1)h}{h}u_n(x) + \frac{nh - t}{h}u_{n-1}(x),
\end{aligned} \tag{3.6}$$

for $t \in ((n-1)h, nh]$, $n = 0, \ldots, N$, see Fig. 3.1.

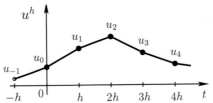
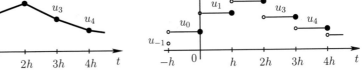

Fig. 3.1 Interpolation of minimizers in time

Since u_n is a minimizer of J_n, the first variation of J_n at u_n vanishes. Thus, for any $\varphi \in H_0^1(\Omega)$ we have

$$
0 = \frac{d}{d\varepsilon} J_n(u_n + \varepsilon\varphi)|_{\varepsilon=0} = \lim_{\varepsilon \to 0} \frac{J_n(u_n + \varepsilon\varphi) - J_n(u_n)}{\varepsilon}
$$

$$
= \lim_{\varepsilon \to 0} \int_\Omega \frac{(2u_n + \varepsilon\varphi - 4u_{n-1} + 2u_{n-2})\varphi}{2h^2} \, dx + \lim_{\varepsilon \to 0} \frac{1}{2} \int_\Omega \left(2\nabla u_n \cdot \nabla \varphi + \varepsilon|\nabla\varphi|^2\right) dx
$$

$$
= \int_\Omega \frac{u_n - 2u_{n-1} + u_{n-2}}{h^2} \varphi \, dx + \int_\Omega \nabla u_n \cdot \nabla\varphi \, dx. \tag{3.7}
$$

Using the definition of $\overline{u}^h$ and u^h in (3.6), this can be rewritten as

$$
\int_\Omega \left[\frac{u_t^h(t) - u_t^h(t-h)}{h} \varphi + \nabla\overline{u}^h(t) \cdot \nabla\varphi \right] dx = 0 \quad \text{for a.e. } t \in (0, T) \ \forall \varphi \in H_0^1(\Omega).
$$

We note that the above relation holds also when multiplied by any $C([0, T])$-function. Hence, integrating over the time interval $(0, T)$ and using a standard density argument, we arrive at

$$
\int_0^T \int_\Omega \left[\frac{u_t^h(t) - u_t^h(t-h)}{h} \varphi + \nabla\overline{u}^h \cdot \nabla\varphi \right] dx \, dt = 0 \qquad \forall \varphi \in L^2(0, T; H_0^1(\Omega)).
$$

$$\tag{3.8}$$

Now, we would like to take the time step h to zero. Thus, our goal is to show that when $h \to 0+$, the functions u^h converge to some function u, which satisfies the following definition.

Definition 3.1 We say that $u \in H^1(Q_T)$ is a weak solution of (3.1)–(3.4) if $u(x, 0) = u_0(x)$ for $x \in \Omega$ and $u(x, t) = 0$ for $x \in \partial\Omega$ in the sense of traces and if

$$
\int_0^T \int_\Omega (-u_t\varphi_t + \nabla u \cdot \nabla\varphi) \, dx \, dt - \int_\Omega v_0\varphi(x, 0) \, dx = 0 \tag{3.9}
$$

holds for any $\varphi \in C_0^\infty(\Omega \times [0, T))$.

To be able to do so, a uniform estimate on the approximate solutions is needed. We state it in the following lemma.

Lemma 3.1 *Suppose Ω is a bounded domain with smooth boundary. Let J_n, $n = 1, 2, \ldots, N$, be the functionals defined by (3.5) and let u_n be the corresponding minimizers in $H_0^1(\Omega)$. Define functions $\overline{u}^h$ and u^h by (3.6) and assume that $h \leq 1$. Then the following estimate holds:*

$$
\|u_t^h(t)\|_{L^2(\Omega)}^2 + \|\nabla\overline{u}^h(t)\|_{L^2(\Omega)}^2 \leq C_E \qquad \text{for a.e. } t \in (0, T), \tag{3.10}
$$

where the constant C_E depends on norms of the initial data u_0, v_0 but is independent of h.

Proof Estimates of this kind are usually derived by testing the equation by the time-derivative of the solution. Setting $\varphi := u_n - u_{n-1}$ in (3.7), we get

$$\int_\Omega \frac{u_n - 2u_{n-1} + u_{n-2}}{h^2}(u_n - u_{n-1})\, dx + \int_\Omega (\nabla u_n - \nabla u_{n-1}) \cdot \nabla u_n\, dx = 0.$$

We employ the inequality

$$\frac{a^2}{2} - \frac{b^2}{2} \le (a-b)a, \qquad \forall a, b \in \mathbb{R}, \tag{3.11}$$

to find that for each $n = 1, 2, \ldots, N$,

$$\int_\Omega \left[\left(\frac{u_n - u_{n-1}}{h} \right)^2 + |\nabla u_n|^2 \right] dx \le \int_\Omega \left[\left(\frac{u_{n-1} - u_{n-2}}{h} \right)^2 + |\nabla u_{n-1}|^2 \right] dx.$$

When these inequalities are summed from $n = 1$ to an arbitrary integer $k \le N$, the terms in between cancel, and we obtain

$$\int_\Omega \left[\left(\frac{u_k - u_{k-1}}{h} \right)^2 + |\nabla u_k|^2 \right] dx \le \int_\Omega \left[\left(\frac{u_0 - u_{-1}}{h} \right)^2 + |\nabla u_0|^2 \right] dx$$

$$= \int_\Omega \left[v_0^2 + |\nabla u_0|^2 \right] dx.$$

This is already the desired estimate (3.10), because $u_t^h(t) = (u_k - u_{k-1})/h$ for $t \in ((k-1)h, kh)$. $\qquad \square$

Thanks to estimate (3.10) and the weak sequential compactness of bounded sets in the reflexive Banach space $L^2(Q_T)$, we can extract a sequence $\{\nabla \overline{u}^{h_k}\}_{k \in \mathbb{N}}$ which converges weakly in $(L^2(Q_T))^m$ to a function v. From the sequence $\{h_k\}_{k \in \mathbb{N}}$ obtained in this way, we can extract another subsequence $\{h_{k_l}\}_{l \in \mathbb{N}}$ so that $\{u_t^{h_{k_l}}\}_{l \in \mathbb{N}}$ converges weakly in $L^2(Q_T)$ to a function U. Having explained the details, we safely omit the lengthy subscripts and simply write

$$\nabla \overline{u}^h \rightharpoonup v \qquad \text{weakly in } (L^2(Q_T))^m, \tag{3.12}$$

$$u_t^h \rightharpoonup U \qquad \text{weakly in } L^2(Q_T). \tag{3.13}$$

We should now show that there is a function $u \in H^1(Q_T)$ such that $v = \nabla u$ and $U = u_t$ in $L^2(Q_T)$. To this end, we prepare a few more estimates.

Lemma 3.2 *Let $\bar{u}^h$ and u^h be defined by (3.6). Then the following relations hold:*

$$\|\bar{u}^h(t) - u^h(t)\|_{L^2(\Omega)} \leq h\|u_t^h(t)\|_{L^2(\Omega)} \qquad \text{for a.e. } t \in (0, T), \tag{3.14}$$

$$\|u^h\|_{L^2(Q_T)}^2 \leq \|\bar{u}^h\|_{L^2(Q_T)}^2 + \frac{h}{2}\|u_0\|_{L^2(\Omega)}^2, \tag{3.15}$$

$$\|\nabla u^h\|_{L^2(Q_T)}^2 \leq \|\nabla\bar{u}^h\|_{L^2(Q_T)}^2 + \frac{h}{2}\|\nabla u_0\|_{L^2(\Omega)}^2. \tag{3.16}$$

Proof First, we estimate the norm of the difference of the approximate functions $\bar{u}^h$ and u^h. Letting $t \in ((n-1)h, nh)$, we have

$$\|\bar{u}^h(t) - u^h(t)\|_{L^2(\Omega)}^2 = \int_\Omega \left(u_n - \frac{t-(n-1)h}{h}u_n - \frac{nh-t}{h}u_{n-1}\right)^2 dx$$

$$= \int_\Omega \left(\frac{nh-t}{h}\right)^2 (u_n - u_{n-1})^2 \, dx$$

$$\leq \int_\Omega (u_n - u_{n-1})^2 \, dx$$

$$= h^2 \int_\Omega (u_t^h(t))^2 \, dx,$$

which proves (3.14).

We have further

$$\|u^h\|_{L^2(Q_T)}^2 - \|\bar{u}^h\|_{L^2(Q_T)}^2 = \int_0^T \int_\Omega \left((u^h)^2 - (\bar{u}^h)^2\right) dx \, dt$$

$$= \sum_{n=1}^N \int_{(n-1)h}^{nh} \int_\Omega \left[\left(\frac{t-(n-1)h}{h}u_n - \frac{nh-t}{h}u_{n-1}\right)^2 - u_n^2\right] dx \, dt$$

$$= \sum_{n=1}^N \int_\Omega \left[-\frac{2h}{3}u_n^2 + \frac{h}{3}u_n u_{n-1} + \frac{h}{3}u_{n-1}^2\right] dx$$

$$\leq \frac{h}{6}\sum_{n=1}^N \int_\Omega \left[-4u_n^2 + u_n^2 + u_{n-1}^2 + 2u_{n-1}^2\right] dx$$

$$= \frac{h}{2}\sum_{n=1}^N \int_\Omega (-u_n^2 + u_{n-1}^2) \, dx$$

$$= \frac{h}{2}\int_\Omega (u_0^2 - u_N^2) \, dx,$$

proving (3.15). The last estimate (3.16) is obtained in a completely analogous way.

$\square$

Now we can specify the limit function.

Proposition 3.1 *There is a common subsequence $\{u^h, \overline{u}^h\}$ and a function $u \in H^1(Q_T)$, such that*

$$\overline{u}^h \to u \qquad \text{strongly in } L^2(Q_T), \tag{3.17}$$

$$\nabla \overline{u}^h \rightharpoonup \nabla u \qquad \text{weakly in } (L^2(Q_T))^m, \tag{3.18}$$

$$u_t^h \rightharpoonup u_t \qquad \text{weakly in } L^2(Q_T). \tag{3.19}$$

Proof From Poincaré's inequality there is a universal constant C_P so that

$$\|u^h\|_{L^2(Q_T)} \leq C_P \|\nabla u^h\|_{L^2(Q_T)} \qquad \text{for all } h \in (0, 1). \tag{3.20}$$

By (3.10), (3.16) and (3.20), u^h is uniformly bounded in $H^1(Q_T)$. Therefore, there is a weakly convergent subsequence in $H^1(Q_T)$ and, by the Rellich–Kondrachov theorem, a strongly converging subsequence in $L^2(Q_T)$ (we always mean "subsequence of the last obtained sequence"). Let us denote the cluster function as u:

$$u^h \rightharpoonup u \qquad \text{weakly in } H^1(Q_T). \tag{3.21}$$

By (3.13), $U = u_t$ holds almost everywhere. Moreover, by (3.12), for any $\varphi \in C_0^\infty(Q_T)$ and any $i = 1, \ldots, m$, we have

$$\int_0^T \int_\Omega \left(\frac{\partial \overline{u}^h}{\partial x_i} - \frac{\partial u^h}{\partial x_i} \right) \varphi \, dx \, dt \to \int_0^T \int_\Omega \left(v_i - \frac{\partial u}{\partial x_i} \right) \varphi \, dx \, dt \qquad \text{as } h \to 0+,$$

while (3.14) implies

$$\int_0^T \int_\Omega \left(\frac{\partial \overline{u}^h}{\partial x_i} - \frac{\partial u^h}{\partial x_i} \right) \varphi \, dx \, dt = -\int_0^T \int_\Omega \left(\overline{u}^h - u^h \right) \frac{\partial \varphi}{\partial x_i} \, dx \, dt \to 0 \quad \text{as } h \to 0+.$$

This means that $v = \nabla u$ almost everywhere in Q_T. $\square$

Now, we can pass to the limit as $h \to 0+$ in (3.8). We shall, for the time being, consider a test function φ belonging to $C_0^\infty(\Omega \times [0, T))$. First, by (3.18) we immediately see that

$$\int_0^T \int_\Omega \nabla \overline{u}^h \cdot \nabla \varphi \, dx \, dt \to \int_0^T \int_\Omega \nabla u \cdot \nabla \varphi \, dx \, dt \qquad \text{as } h \to 0+. \tag{3.22}$$

In addition, we have

$$\int_0^T \int_\Omega \frac{u_t^h(t) - u_t^h(t-h)}{h} \varphi \, dx \, dt$$

$$= \int_0^T \int_\Omega \frac{u_t^h(t)}{h} \varphi(t) \, dx \, dt - \int_{-h}^{T-h} \int_\Omega \frac{u_t^h(t)}{h} \varphi(t+h) \, dx \, dt$$

$$= - \int_0^T \int_\Omega u_t^h(t) \frac{\varphi(t+h) - \varphi(t)}{h} \, dx \, dt \qquad (3.23)$$

$$- \int_{-h}^0 \int_\Omega \frac{u_t^h(t)}{h} \varphi(t+h) \, dx \, dt + \int_{T-h}^T \int_\Omega \frac{u_t^h(t)}{h} \varphi(t+h) \, dx \, dt$$

$$\to - \int_0^T \int_\Omega u_t \varphi_t \, dx \, dt - \int_\Omega v_0 \varphi(0) \, dx \qquad \text{as } h \to 0+ .$$

The convergence in the last line is deduced from the following facts: in the first term of (3.23), u_t^h converges weakly and $(\varphi(t) - \varphi(t+h))/h$ converges strongly in $L^2(Q_T)$; in the second term, $u_t^h = (u_0 - u_{-1})/h = v_0$ for $t \in (-h, 0)$; in the third term, $\varphi(t+h) = 0$ for $t \in (T-h, T)$. Thus, we can finally state that

$$\int_0^T \int_\Omega (-u_t \varphi_t + \nabla u \cdot \nabla \varphi) \, dx \, dt - \int_\Omega v_0 \varphi(0, x) \, dx = 0 \quad \forall \varphi \in C_0^\infty(\Omega \times [0, T)).$$

$$(3.24)$$

Noting that the space of functions from $H^1(Q_T)$ with zero trace on $(\Omega \times \{0\}) \cup (\partial\Omega \times [0, T])$ is a closed linear subspace of $H^1(Q_T)$ and, therefore, weakly closed by Mazur's theorem, we conclude by (3.21) that u belongs to this space. Consequently, u satisfies boundary condition (3.2) and initial condition (3.3) in the sense of traces. We remark that the convergence of traces follows also from the compactness of the trace operator $T : H^1(\Omega) \to L^2(\partial\Omega)$. Moreover, u, as a function from $H^1(0, T; L^2(\Omega))$, belongs to $C([0, T]; L^2(\Omega))$. Thus, the initial condition (3.3) is satisfied even in the strong sense.

To summarize, by applying the discrete Morse flow method we have proved that there exists a weak solution $u \in H^1(Q_T)$ to problem (3.1)–(3.4) in the sense of (3.24), satisfying boundary and initial conditions (3.2), (3.3) in the sense of traces.

3.2 Energy-Preserving Scheme for the Linear Wave Equation

The semi-discrete scheme presented in the previous section has one serious drawback—it does not preserve energy, although energy preservation is a key feature of the wave equation. Indeed, the energy estimate proved in Lemma 3.1 reads

$$\|u_t^h(t)\|_{L^2(\Omega)}^2 + \|\nabla \overline{u}^h(t)\|_{L^2(\Omega)}^2 \leq \|v_0\|_{L^2(\Omega)}^2 + \|\nabla u_0\|_{L^2(\Omega)}^2 \qquad \text{for a.e. } t \in (0, T),$$

$$(3.25)$$

which says that the energy of the approximate solution does not increase as time passes. In fact, numerical experiments show that the energy strictly decreases, and this is detrimental not only to the quality of numerical results but also to the development of the theory.

Here we present a modification of the scheme from Sect. 3.1, which guarantees energy preservation for approximate solutions. We shall solve the same problem for the linear wave equation as in Sect. 3.1, that is, we consider a bounded domain $\Omega \subset \mathbb{R}^m$ with smooth boundary $\partial\Omega$, on which the homogeneous Dirichlet boundary condition is given, and a time interval $(0, T)$, $T > 0$. Given an initial position $u_0 \in H_0^1(\Omega)$ and an initial velocity $v_0 \in H_0^1(\Omega)$, we search for $u \in H^1(Q_T)$ satisfying

$$u_{tt}(x, t) = \Delta u(x, t) \qquad \text{in } Q_T := \Omega \times (0, T), \qquad (3.26)$$

$$u(x, t) = 0 \qquad \text{on } \partial\Omega \times (0, T), \qquad (3.27)$$

$$u(x, 0) = u_0(x) \qquad \text{in } \Omega, \qquad (3.28)$$

$$u_t(x, 0) = v_0(x) \qquad \text{in } \Omega. \qquad (3.29)$$

Our aim is to construct a weak solution to this problem, which we recall is given in Definition 3.1. The key idea is to inductively minimize the functional

$$J_n^{CN}(u) := \int_\Omega \frac{|u - 2u_{n-1} + u_{n-2}|^2}{2h^2} \, dx + \frac{1}{4} \int_\Omega |\nabla u + \nabla u_{n-2}|^2 \, dx, \qquad n = 1, 2, \ldots,$$

over $\mathcal{K} := H_0^1(\Omega)$, where the first two functions u_{-1}, u_0 are given, i.e., we set $u_{-1} := u_0 - hv_0$. As in the standard discrete Morse flow, we have the following result.

Proposition 3.2 *Functionals* J_n^{CN} *have the following properties:*

(i)   (Existence of minimizers) *There is a unique minimizer of* J_n^{CN} *in* $\mathcal{K}$.

(ii) (First variation formula) *Any minimizer* u_n *of* J_n^{CN} *satisfies*

$$\int_\Omega \left[\frac{u_n - 2u_{n-1} + u_{n-2}}{h^2} \phi + \nabla\left(\frac{u_n + u_{n-2}}{2}\right) \cdot \nabla\phi \right] dx = 0$$

for any $\phi \in H_0^1(\Omega)$.

Proof

(i)   Since the functionals J_n^{CN} are convex, lower semicontinuous and coercive, we can apply the direct method as usual.

(ii) For a minimizer u_n of J_n^{CN}, we have

$$\frac{d}{d\varepsilon} J_n^{CN}(u_n + \varepsilon\phi)\bigg|_{\varepsilon=0} = 0.$$

Direct calculation concludes the proof.   □

The second property shows that u_n solves the discretized modified wave equation

$$\frac{u - 2u_{n-1} + u_{n-2}}{h^2} = \Delta\left(\frac{u + u_{n-2}}{2}\right)$$

in a weak sense. Based on this fact, we refer to the scheme as a 'Crank–Nicolson' type scheme, and use the notation J_n^{CN} for the functionals. Unlike the standard discrete Morse flow method which uses the functional (3.5), we have the following fine property.

Theorem 3.1 (Energy Conservation)
The quantity

$$E_k := \frac{1}{2}\left\| \frac{u_k - u_{k-1}}{h} \right\|^2_{L^2(\Omega)} + \frac{1}{4}\left(\|\nabla u_k\|^2_{L^2(\Omega)} + \|\nabla u_{k-1}\|^2_{L^2(\Omega)} \right) \tag{3.30}$$

is independent of $k \geq 0$.

Proof For $n = 2, \ldots, k$, since

$$0 = \frac{d}{d\theta} J_n^{CN}(u_n + \theta(u_{n-2} - u_n))\bigg|_{\theta=0},$$

after simple calculation, we get

$$0 = \int_\Omega \left[\frac{(u_{n-1} - u_{n-2})^2 - (u_n - u_{n-1})^2}{h^2} + \frac{1}{2}|\nabla u_{n-2}|^2 - \frac{1}{2}|\nabla u_n|^2 \right] dx.$$

Summing over $n = 1, \ldots, k$, we arrive at

$$\int_\Omega \left[\frac{1}{h^2}(u_0 - u_{-1})^2 - \frac{1}{h^2}(u_k - u_{k-1})^2 \right.$$
$$\left. + \frac{1}{2}|\nabla u_{-1}|^2 + \frac{1}{2}|\nabla u_0|^2 - \frac{1}{2}|\nabla u_{k-1}|^2 - \frac{1}{2}|\nabla u_k|^2 \right] dx = 0,$$

which implies

$$E_0 = \frac{1}{2} \int_\Omega \left[\frac{1}{h^2}(u_0 - u_{-1})^2 + \frac{1}{2}|\nabla u_{-1}|^2 + \frac{1}{2}|\nabla u_0|^2 \right] dx$$
$$= \frac{1}{2} \int_\Omega \left[\frac{1}{h^2}(u_k - u_{k-1})^2 + \frac{1}{2}|\nabla u_{k-1}|^2 + \frac{1}{2}|\nabla u_k|^2 \right] dx$$
$$= E_k,$$

and the proof is complete.   □

The quantity E_k corresponds to a time-discretization of the energy

$$\frac{1}{2} \int_\Omega (u_t)^2 \, dx + \frac{1}{2} \int_\Omega |\nabla u|^2 \, dx.$$

Thus, the above theorem tells us that energy is conserved at the discrete level, which is essential for the accuracy of numerical computations.

The construction of a weak solution follows the same procedure as in Sect. 3.1 above. We outline the main steps here and provide more details later in the setting of a hyperbolic free boundary problem (see Chap. 5).

Time Interpolation

As in the standard discrete Morse flow, we interpolate minimizers in time. For $n = 0, 1, \ldots, N$ and $(x, t) \in \Omega \times ((n-1)h, nh]$, we define

$$\bar{u}^h(x, t) = u_n(x),$$
$$u^h(x, t) = \frac{t - (n-1)h}{h} u_n(x) + \frac{nh - t}{h} u_{n-1}(x).$$

Approximate Weak Solution

Rewriting the formula for the first variation in Proposition 3.1 (ii) in terms of $\bar{u}^h$, u^h, we have, for any $\varphi \in L^2(0, T; H_0^1(\Omega))$,

$$\int_h^T \int_\Omega \left(\frac{u_t^h(t) - u_t^h(t-h)}{h} \varphi + \nabla \frac{\bar{u}^h(t) + \bar{u}^h(t-2h)}{2} \cdot \nabla\varphi \right) dx\, dt = 0.$$

$$(3.31)$$

If we can pass to the limit as $h \to 0+$ in (3.31), we expect to obtain the formula (3.9) in the definition of a weak solution.

Passing to the Limit as $h \to 0+$

As a corollary of the energy conservation (Theorem 3.1), we obtain the energy estimate

$$\|u_t^h(t)\|_{L^2(\Omega)}^2 + \|\nabla\bar{u}^h(t)\|_{L^2(\Omega)}^2 \le C(u_0, v_0)$$

for a.e. $t \in (0, T)$. Hence there is $u \in H^1(Q_T)$ and a decreasing sequence $\{h_j\}$ with $h_j \downarrow 0+$ (which we do not relabel) such that

$$\begin{cases} \nabla\bar{u}^h \rightharpoonup \nabla u & \text{weakly in } L^2(Q_T), \\ u_t^h \rightharpoonup u_t & \text{weakly in } L^2(Q_T), \\ u^h \to u & \text{strongly in } L^2(Q_T). \end{cases}$$

Moreover, a direct calculation shows

$$\nabla\bar{u}^h(\cdot - 2h) \rightharpoonup \nabla u \quad \text{weakly in } L^2(Q_T).$$

This information already allows us to take $h \to 0+$ in (3.31).

3.3 Volume-Constrained Hyperbolic Problem

There are several other methods of proving the well-posedness of initial-boundary value problems for the linear wave equation, and thus it is not clear whether it is meaningful to try to develop a new one. However, it turns out that the discrete Morse flow method is a powerful tool capable of handling a wide range of nonlinear hyperbolic problems. We present two important examples of such applications, starting with a hyperbolic problem with a nonlocal constraint, and continuing with free boundary problems in the following chapter.

In particular, we shall study an initial-boundary value problem for the wave equation with a volume constraint. Let $Q_T = \Omega \times (0, T)$, where $T > 0$ and Ω is a bounded domain in $\mathbb{R}^m$ with Lipschitz boundary $\partial\Omega$. Further, let V be a positive real number, and $u_0, v_0 \in H^1(\Omega)$ be the initial data, which we require to satisfy the compatibility conditions $u_0(x) = v_0(x) = 0$ for $x \in \partial\Omega$ and $\int_\Omega u_0(x)\,dx = V$. Then the problem is to find a function $u : Q_T \to \mathbb{R}$ fulfilling

$$u_{tt}(x, t) = \Delta u(x, t) + \lambda(t) \qquad \text{in } Q_T, \tag{3.32}$$

$$u(x, t) = 0 \qquad \text{on } \partial\Omega \times (0, T), \tag{3.33}$$

$$u(x, 0) = u_0(x) \qquad \text{in } \Omega, \tag{3.34}$$

$$u_t(x, 0) = v_0(x) \qquad \text{in } \Omega, \tag{3.35}$$

where λ is determined by the volume-preserving condition

$$\int_\Omega u(x, t)\,dx = V \qquad \forall t \in (0, T), \tag{3.36}$$

and is assumed to take the form

$$\lambda(t) = \frac{1}{V} \int_\Omega (u_{tt}u + |\nabla u|^2)\,dx. \tag{3.37}$$

Following the procedure of discrete Morse flow, we divide the time interval $(0, T)$ into N parts and set the time-step $h = T/N$. The discretized functional reads

$$J_n(u) = \int_\Omega \frac{|u - 2u_{n-1} + u_{n-2}|^2}{2h^2}\,dx + \frac{1}{2}\int_\Omega |\nabla u|^2\,dx. \tag{3.38}$$

Defining u_0 and $u_{-1} := u_0 - hv_0$, for $n = 1, \ldots, N$, we inductively search for a minimizer u_n of $J_n(u)$ in the convex set

$$\mathcal{K}_{\text{vol}} = \left\{ u \in H^1(\Omega); \ u|_{\partial\Omega} = 0, \int_\Omega u\,dx = V \right\}.$$

In order to derive a weak formulation, we apply the method of Lagrange multipliers. We choose a function $\zeta \in C_0^\infty(\Omega)$ and write down the condition that the first variation of the functional $J_n^\lambda(u) = J_n(u) - \lambda_n \int_\Omega u\,dx$ vanishes. This yields

$$\int_\Omega \frac{u - 2u_{n-1} + u_{n-2}}{h^2}\zeta\,dx + \int_\Omega \nabla u \cdot \nabla \zeta\,dx = \int_\Omega \lambda_n \zeta\,dx, \tag{3.39}$$

and holds also for $\zeta \in H_0^1(\Omega)$. Setting $\zeta = u$ we get a specific form of the multiplier as

$$\lambda_n = \frac{1}{V} \int_\Omega \left(\frac{u - 2u_{n-1} + u_{n-2}}{h^2} u + |\nabla u|^2 \right) dx. \tag{3.40}$$

This form of Lagrange multiplier can also be computed using volume-preserving perturbations

$$u^\varepsilon = V \frac{u + \varepsilon \zeta}{\int_\Omega (u + \varepsilon \zeta) \, dx} \in \mathcal{K}_{vol},$$

which yields

$$0 = \frac{d}{d\varepsilon} J_n(u^\varepsilon)|_{\varepsilon=0}$$

$$= \int_\Omega \left[\left(\frac{u - 2u_{n-1} + u_{n-2}}{h^2} \zeta + \nabla u \cdot \nabla \zeta \right) \right.$$

$$\left. - \frac{1}{V} \left(\int_\Omega \zeta \, dx \right) \left(\frac{u - 2u_{n-1} + u_{n-2}}{h^2} u + |\nabla u|^2 \right) \right] dx,$$

and again leads to the formulas (3.39), (3.40). We remark that the function u satisfying the above Euler–Lagrange equations is denoted by u_n and thus we may write

$$\int_\Omega \frac{u_n - 2u_{n-1} + u_{n-2}}{h^2} \zeta \, dx + \int_\Omega \nabla u_n \cdot \nabla \zeta \, dx = \int_\Omega \lambda_n \zeta \, dx. \tag{3.41}$$

We interpolate the obtained minimizers in time and define the functions u^h, $\overline{u}^h$ as in (3.6). In addition, we interpolate the discrete Lagrange multiplier: in $\Omega \times ((n - 1)h, nh]$ we set $\overline{\lambda}^h(t) = \lambda_n$, $n = 1, \ldots, N$. Then, by (3.39), we see that for any $\phi \in L^2(0, T; H_0^1(\Omega))$ the following identity holds:

$$\int_0^T \int_\Omega \left(\frac{u_t^h(t) - u_t^h(t - h)}{h} \phi + \nabla \overline{u}^h \cdot \nabla \phi \right) dx \, dt = \int_0^T \int_\Omega \overline{\lambda}^h \phi \, dx \, dt. \tag{3.42}$$

It is natural to call u^h and $\overline{u}^h$ defined by the sequence of minimizers $\{u_n\}$ an *approximate weak solution* of (3.32)–(3.35). As the next step, we would like to show that the approximate solutions defined above converge, as $h \to 0+$, to a weak solution of the original problem, defined as follows:

Definition 3.2 A function $u \in H^1(0, T; L^2(\Omega)) \cap L^\infty(0, T; H_0^1(\Omega))$ is called a weak solution to (3.32)–(3.35) if $u(0, \cdot) = u_0$ is satisfied in the sense of traces and the following identity holds for all $\phi \in C_0^\infty(\Omega \times [0, T))$:

$$\int_0^T \int_\Omega (-u_t \phi_t + \nabla u \cdot \nabla \phi) \, dx \, dt - \int_\Omega v_0 \phi(0) \, dx$$

$$= \frac{1}{V} \int_0^T \int_\Omega \left(-u_t(u\Phi)_t + |\nabla u|^2 \Phi\right) dx \, dt - \frac{1}{V} \int_\Omega u_0 v_0 \Phi(0) \, dx.$$

Here we have written $\Phi(t) = \int_\Omega \phi(x, t) \, dx$.

Note that the right-hand side actually stands for the expression $\int_0^T \int_\Omega \lambda \phi \, dx \, dt$ with λ from (3.37) but it is integrated by parts with respect to time.

The stepping stone to the proof of the convergence is the following energy estimate.

Proposition 3.3 *There is a constant C_E independent of h such that*

$$\|u_t^h(t)\|_{L^2(\Omega)}^2 + \|\nabla \overline{u}^h(t)\|_{L^2(\Omega)}^2 \le C_E \qquad \forall t \in (0, T). \tag{3.43}$$

Proof We select the test function $\psi = (1 - \theta)u_n + \theta u_{n-1}$, $\theta \in (0, 1)$. Noting that $\psi \in \mathcal{K}_{\text{vol}}$, by the minimality property we get

$$0 \le \frac{1}{\theta}(J_n(\psi) - J_n(u_n))$$

$$= \frac{1}{2h^2} \int_\Omega \left(2(u_{n-1} - u_n)(u_n - 2u_{n-1} + u_{n-2}) + \theta(u_n - u_{n-1})^2\right) dx$$

$$+ \frac{1}{2} \int_\Omega \left(2\nabla u_n \cdot \nabla(u_{n-1} - u_n) + \theta|\nabla(u_{n-1} - u_n)|^2\right) dx.$$

Passing to the limit as $\theta \to 0+$, we find that

$$0 \le -\frac{1}{h^2} \int_\Omega (u_n - u_{n-1})(u_n - 2u_{n-1} + u_{n-2}) \, dx + \int_\Omega \nabla u_n \cdot \nabla(u_{n-1} - u_n) \, dx$$

$$\le \frac{1}{2h^2} \int_\Omega \left((u_{n-1} - u_{n-2})^2 - (u_n - u_{n-1})^2\right) dx + \frac{1}{2} \int_\Omega \left(|\nabla u_{n-1}|^2 - |\nabla u_n|^2\right) dx.$$

Thus, after summing over $n = 1, \ldots, k$, we arrive at

$$\frac{1}{h^2}\|u_k - u_{k-1}\|_{L^2(\Omega)}^2 + \|\nabla u_k\|_{L^2(\Omega)}^2 \le \frac{1}{h^2}\|u_0 - u_{-1}\|_{L^2(\Omega)}^2 + \|\nabla u_0\|_{L^2(\Omega)}^2.$$

Hence we can set $C_E = \|v_0\|_{L^2(\Omega)}^2 + \|\nabla u_0\|_{L^2(\Omega)}^2$, and the proposition is proved.
$\qquad\qquad\qquad\qquad\qquad\qquad\qquad\qquad\qquad\qquad\qquad\qquad\qquad\qquad\qquad\qquad\square$

Consequently, there is a subsequence of $u^h, \bar{u}^h$ such that Proposition 3.1 holds true. If one assumes higher regularity of the initial data, the following uniform estimate is obtained.

Lemma 3.3 *Let $u_0 \in H^2(\Omega)$. Then the approximate weak solution obeys the estimate*

$$\int_\Omega \left| \frac{u_t^h(t) - u_t^h(t-h)}{h} \right|^2 dx + \int_\Omega |\nabla u_t^h(t)|^2\, dx \le C_E' \qquad \forall t \in (0, T), \qquad (3.44)$$

where the constant C_E' is independent of h.

Proof Let us set $u_{-2} = u_{-1} - h v_0 + h^2 \Delta u_0$, where u_{-1}, u_0 were defined after (3.38) using the initial conditions (3.34), (3.35). This function may not satisfy the volume constraint but since

$$\frac{u_0 - 2u_{-1} + u_{-2}}{h^2} = \Delta u_0,$$

we have for every $\zeta \in H_0^1(\Omega)$

$$\int_\Omega \frac{u_0 - 2u_{-1} + u_{-2}}{h^2} \zeta\, dx + \int_\Omega \nabla u_0 \cdot \nabla \zeta\, dx = \int_\Omega \lambda_0 \zeta\, dx = 0.$$

Indeed,

$$\lambda_0 = \frac{1}{V} \int_\Omega \left(\frac{u_0 - 2u_{-1} + u_{-2}}{h^2} u_0 + |\nabla u_0|^2 \right) dx$$

$$= \frac{1}{V} \int_\Omega (\Delta u_0\, u_0 + |\nabla u_0|^2)\, dx = \frac{1}{V} \int_\Omega (-|\nabla u_0|^2 + |\nabla u_0|^2)\, dx = 0.$$

This means that u_0 is a minimizer of J_0 in $\mathcal{K}_{\mathrm{vol}}$. Using the notation $d_n = u_n - u_{n-1}$, $n = -1, 0, \ldots, N$, we subtract Eq. (3.41), where we write $n-1$ instead of n from the equation itself. We get

$$\int_\Omega \frac{d_n - 2d_{n-1} + d_{n-2}}{h^2} \zeta\, dx + \int_\Omega \nabla d_n \cdot \nabla \zeta\, dx = \int_\Omega (\lambda_n - \lambda_{n-1}) \zeta\, dx, \qquad n = 1, 2, \ldots, N.$$

For $n = 1, 2, \ldots, N$ we choose $\zeta = d_n - d_{n-1}$ and since this function has zero volume, we find that for every $n = 1, 2, \ldots, N$,

$$\int_\Omega \frac{d_n - 2d_{n-1} + d_{n-2}}{h^2} (d_n - d_{n-1})\, dx + \int_\Omega \nabla d_n \cdot \nabla (d_n - d_{n-1})\, dx = 0.$$

Estimating and summing up in the fashion of the proof of Proposition 3.3 yields

$$\frac{1}{h^2} \int_\Omega (d_n - d_{n-1})^2 \, dx + \int_\Omega |\nabla d_n|^2 \, dx \le \frac{1}{h^2} \int_\Omega (d_0 - d_{-1})^2 \, dx + \int_\Omega |\nabla d_0|^2 \, dx,$$

for $n = 1, 2, \ldots, N$, which is the same as

$$\int_\Omega \left| \frac{u_t^h(t) - u_t^h(t-h)}{h} \right|^2 \, dx + \int_\Omega |\nabla u_t^h(t)|^2 \, dx \le \int_\Omega |\Delta u_0|^2 \, dx + \int_\Omega |\nabla v_0|^2 \, dx$$

for $t \in (0, T)$.   □

This lemma tells us that there exists $u_{tt} \in L^2(Q_T)$ and a further subsequence to that of Proposition 3.1 such that

$$\frac{u_t^h(t) - u_t^h(t-h)}{h} \rightharpoonup u_{tt} \quad \text{weakly in } L^2(Q_T). \tag{3.45}$$

Moreover, it is immediate by Proposition 3.1 that the limit function has zero trace on $\partial\Omega$ and satisfies the volume constraint (3.36).

Theorem 3.2 *The limit function u obtained above is a weak solution of the original problem* (3.32)–(3.35).

Proof Let $\phi \in C_0^\infty(\Omega \times [0, T))$ and define as always $\Phi(t) = \int_\Omega \phi(x, t) \, dx$. Then

$$\int_0^T \int_\Omega \nabla \overline{u}^h \cdot \nabla \phi \, dx \, dt \to \int_0^T \int_\Omega \nabla u \cdot \nabla \phi \, dx \, dt \qquad \text{as } h \to 0+. \tag{3.46}$$

Moreover, we have

$$\int_0^T \int_\Omega \frac{u_t^h(t) - u_t^h(t-h)}{h} \phi \, dx \, dt$$

$$= \int_0^T \int_\Omega \frac{u_t^h(t)}{h} \phi(t) \, dx \, dt - \int_{-h}^{T-h} \int_\Omega \frac{u_t^h(t)}{h} \phi(t+h) \, dx \, dt$$

$$= -\int_0^T \int_\Omega u_t^h(t) \frac{\phi(t+h) - \phi(t)}{h} \, dx \, dt - \int_{-h}^0 \int_\Omega \frac{u_t^h(t)}{h} \phi(t+h) \, dx \, dt$$

$$+ \int_{T-h}^T \int_\Omega \frac{u_t^h(t)}{h} \phi(t+h) \, dx \, dt$$

$$\to -\int_0^T \int_\Omega u_t \phi_t \, dx \, dt - \int_\Omega v_0 \phi(x, 0) \, dx \qquad \text{as } h \to 0+. \tag{3.47}$$

It remains to pass to the limit in the right-hand side of (3.42). To do so, let $\vartheta \in C_0^\infty(\Omega \times [0, T))$ and $\Theta(t) = \int_\Omega \vartheta \, dx$, and let v be any smooth function independent of t, satisfying boundary conditions and the volume constraint. Then setting $\phi = (\bar{u}^h - v)\Theta \in L^2(0, T; H_0^1(\Omega))$ in (3.42), we get

$$\int_0^T \int_\Omega \left(\frac{u_t^h(t) - u_t^h(t-h)}{h} \bar{u}^h \Theta + |\nabla \bar{u}^h|^2 \Theta \right) dx\, dt$$

$$= \int_0^T \int_\Omega \left(\frac{u_t^h(t) - u_t^h(t-h)}{h} v\Theta + \nabla \bar{u}^h \cdot \nabla v\Theta \right) dx\, dt.$$

Employing this identity we have

$$\int_0^T \int_\Omega \bar{\lambda}^h \vartheta \, dx\, dt = \int_0^T \bar{\lambda}^h \Theta \, dt$$

$$= \frac{1}{V} \int_0^T \int_\Omega \left(\frac{u_t^h(t) - u_t^h(t-h)}{h} \bar{u}^h \Theta + |\nabla \bar{u}^h|^2 \Theta \right) dx\, dt$$

$$= \frac{1}{V} \int_0^T \int_\Omega \left(\frac{u_t^h(t) - u_t^h(t-h)}{h} v\Theta + \nabla \bar{u}^h \cdot \nabla v\Theta \right) dx\, dt.$$

Now we can take $h \to 0+$. By calculations similar to (3.47) (where we write $v\Theta$ instead of ϕ), we obtain

$$\int_0^T \bar{\lambda}^h \Theta \, dt$$

$$\to \frac{1}{V} \left(\int_0^T \int_\Omega -u_t v\Theta_t \, dx\, dt - \int_\Omega u_t(0) v\Theta(0) \, dx + \int_0^T \int_\Omega \nabla u \cdot \nabla v\Theta \, dx\, dt \right).$$

Passing to the limit in the approximate weak equation (3.42) we arrive at

$$\int_0^T \int_\Omega (-u_t \vartheta_t + \nabla u \cdot \nabla \vartheta) \, dx\, dt - \int_\Omega u_t(0)\vartheta(0) \, dx \tag{3.48}$$

$$= \frac{1}{V} \left(\int_0^T \int_\Omega -u_t v\Theta_t \, dx\, dt - \int_\Omega u_t(0) v\Theta(0) \, dx + \int_0^T \int_\Omega \nabla u \cdot \nabla v\Theta \, dx\, dt \right).$$

In the last identity, we do the same trick as above and choose $\vartheta = (u - v)\Phi$, which is possible due to a density argument, arriving at

$$\int_0^T \int_\Omega \left(-u_t(u\Phi)_t + |\nabla u|^2 \Phi \right) dx\, dt - \int_\Omega u_t(0)u(0)\Phi(0) \, dx \tag{3.49}$$

$$= \int_0^T \int_\Omega (-u_t v\Phi_t + \nabla u \cdot \nabla v\Phi) \, dx\, dt - \int_\Omega u_t(0)v\Phi(0) \, dx.$$

We have used the fact that the limit function u satisfies the volume condition which is shown by the strong convergence of $\overline{u}^h(t)$ in $L^2(\Omega)$.

Now using the original test function ϕ instead of ϑ in (3.48) and combining with (3.49), we obtain

$$\int_0^T \int_\Omega (-u_t \phi_t + \nabla u \cdot \nabla \phi) \, dx \, dt - \int_\Omega u_t(0)\phi(0) \, dx$$

$$= \frac{1}{V} \int_0^T \int_\Omega \left(-u_t(u\Phi)_t + |\nabla u|^2 \Phi\right) dx \, dt - \frac{1}{V} \int_\Omega u_t(0)u(0)\Phi(0) \, dx,$$

which is exactly the identity from the definition of a weak solution.   □

Chapter 4
Discrete Morse Flow with Free Boundary

In this chapter we analyze the discrete Morse flow approximation applied to the hyperbolic free boundary problem of finding stationary points of the functional (2.11), i.e.,

$$I_\varepsilon(u) = \int_0^T \int_\Omega \left((u_t)^2 \chi_{\{u>0\}} - |\nabla u|^2 - \Gamma B_\varepsilon(u) \right) dx \, dt. \qquad (4.1)$$

Several more or less similar variants of the discrete Morse flow approximation have been proposed. The pioneering work of H. Yoshiuchi, S. Omata, K. Svadlenka and K. Ohara [92] was inspired by the work of A. Tachikawa [86]. Later, K. Kikuchi suggested a method employing indicator functions [34], and a still different approach was developed in [83]. We shall give a detailed account of the last method in Sect. 4.2, and of the first method in connection with the energy-preserving scheme in Sect. 4.3, while we only briefly touch on the approach using indicator functions. We note that the elliptic case was scrutinized in [58].

4.1 Weak Solution and Discrete Morse Flow

The functional (4.1) has been derived to describe the tape peeling phenomenon but to present the analysis of discrete Morse flow approximation, we do not have to use this particular form. Therefore, we shall address the following form of functionals:

$$I(u) = \int_0^T \int_\Omega \left(\frac{1}{2}(u_t)^2 \chi_{\{u>0\}} - \frac{1}{2}|\nabla u|^2 - F(u) \right) dx \, dt. \qquad (4.2)$$

In particular, we have replaced the regularized adhesion force term $\Gamma B_\varepsilon(u)$ by a general outer force $F(x, t, u)$, which may depend not only on the unknown function

© The Author(s), under exclusive license to Springer Nature Singapore Pte Ltd. 2022
S. Omata et al., *Variational Approach to Hyperbolic Free Boundary Problems*,
SpringerBriefs in Mathematics, https://doi.org/10.1007/978-981-19-6731-3_4

u but also directly on x and t. In the sequel we shall suppose that the function F and its derivative $f(x,t,u) = (\partial/\partial u)F(x,t,u)$ satisfy the following conditions:

(F1) $f(x,t,u)$ is measurable in x, differentiable in t and continuous in u.
(F2) $F(x,t,u) = 0$ whenever $u \leq 0$.
(F3) f satisfies the bound

$$|f(x,t,u)| \leq 2C_f|u| + \vartheta(x,t), \qquad \vartheta \in L^\infty(Q_T) \qquad (4.3)$$

with $\vartheta \geq 0$ and

$$0 \leq C_f \leq \min\left\{ \frac{1}{4C_P^2}, \frac{1}{2\sqrt{2}C_P} \right\}.$$

Here C_P is a constant depending only on the dimension m and on $|\Omega|$, which is derived from Poincaré's inequality:

$$\|u\|_{L^2(\Omega)} \leq C_P \|\nabla u\|_{L^2(\Omega)} \qquad \text{for all } u \in H_0^1(\Omega).$$

We recall that Ω is a bounded domain in $\mathbb{R}^m$ with sufficiently smooth, for example piecewise Lipschitz, boundary. Note that the primitive function F can be chosen such that

$$|F(x,t,u)| \leq C_f u^2 + \vartheta(x,t)|u|.$$

Several remarks are in order. First, we have generalized the outer force term in the functional but still we have in mind mainly the regularized adhesion force, which is consistent because the function B_ε fulfills all the above assumptions. The far goal then is to take the regularizing parameter to zero. Of course, the above assumptions on F could be weakened at the price of a more complicated exposition in what follows. Moreover, a regularization of the adhesion force or its replacement by another smooth term do not alter the fact that the problem of finding stationary points of I leads to a hyperbolic problem with free boundary, due to the presence of characteristic function in the kinetic energy term (and possibly due to volume constraint). Finally, the term $|\nabla u|^2$ leading to the Laplace operator in the corresponding Euler–Lagrange equation could be also replaced by a general elliptic operator, but again, we will favor simplicity over completeness.

We proceed with considerations about the set of admissible functions in which stationary points of I will be searched for. In the sequel, we will deal with two types of problems: a problem of finding stationary points of I in the set

$$\mathcal{K}^T = \left\{ u \in H^1(Q_T);\ u(x,t) = 0 \text{ for } (x,t) \in \partial\Omega \times (0,T), \right.$$

$$\left. u(x,0) = u_0(x),\ u_t(x,0) = v_0(x) \text{ for } x \in \Omega \right\}, \qquad (4.4)$$

where the equalities are considered in an appropriate weak sense, and a volume-constrained problem, where the admissible set has the form

$$\mathcal{K}_V^T = \left\{ u \subset \mathcal{K}^T; \ \int_\Omega u(x, t) \, dx = V \text{ for a.e. } t \in (0, T) \right\}.$$

Again, one could, under suitable assumptions, generalize the homogeneous Dirichlet boundary condition and the form of the constraint in $\mathcal{K}_V^T$.

In the following sections, we will construct a sequence of approximate solutions, and, under additional assumptions, prove the existence of a weak solution to the problem of finding stationary points of the functional I in the sets $\mathcal{K}^T$ or $\mathcal{K}_V^T$. This weak solution to the basic problem without constraint is defined as follows.

Definition 4.1 A weak solution is a function $u \in H^1\left(0, T; L^2(\Omega)\right) \cap L^\infty\left(0, T; H_0^1(\Omega)\right)$ satisfying the following equality for all test functions $\phi \in C_0^\infty(\Omega \times [0, T) \cap \{u > 0\})$:

$$\int_0^T \int_\Omega \left(-u_t \phi_t + \nabla u \cdot \nabla \phi + F'(u)\phi\right) dx \, dt - \int_\Omega v_0 \phi(x, 0) \, dx = 0. \qquad (4.5)$$

Moreover, we require that $u \equiv 0$ is satisfied outside of $\{u > 0\}$, and that u fulfills the initial condition $u(0, x) = u_0(x)$ and boundary condition in the sense of traces.

Let us define the weak solution to the volume-constrained problem, too, following Definition 3.2.

Definition 4.2 A weak solution to the volume-constrained problem is a function $u \in H^1\left(0, T; L^2(\Omega)\right) \cap L^\infty\left(0, T; H_0^1(\Omega)\right)$ satisfying the following equality for all test functions $\phi \in C_0^\infty(\Omega \times [0, T) \cap \{u > 0\})$:

$$\int_0^T \int_\Omega \left(-u_t \phi_t + \nabla u \cdot \nabla \phi + F'(u)\phi\right) dx \, dt - \int_\Omega v_0 \phi(0, x) \, dx \qquad (4.6)$$

$$= \frac{1}{V} \int_0^T \int_\Omega \left(-u_t (u\Phi)_t + |\nabla u|^2 \Phi + F'(u)u\Phi\right) dx \, dt - \frac{1}{V} \int_\Omega u_0 v_0 \Phi(0) \, dx,$$

where $\Phi(t) = \int_\Omega \phi(x, t) \, dx$. Moreover, we require that $u \equiv 0$ is satisfied outside of $\{u > 0\}$, and that u fulfills the initial condition $u(0, x) = u_0(x)$ and boundary condition in the sense of traces.

We will challenge the problem of existence of weak solutions armed with the discrete Morse flow method. Accordingly, for a given $N \in \mathbb{N}$ we define the time step $h = T/N$. Setting the stationary version of the admissible set $\mathcal{K}^T$ as

$$\mathcal{K} = H_0^1(\Omega),$$

we take the initial condition $u_0 \in \mathcal{K}$ and for $v_0 \in H_0^1(\Omega)$ define $u_{-1} \in \mathcal{K}$ by $u_{-1} = u_0 - h v_0$. The discrete Morse flow then amounts to minimizing the functionals

$$J_n(u) = \mathcal{M}(u, u_{n-1}, u_{n-2}) + \int_\Omega \left(\frac{1}{2} |\nabla u|^2 + F(u) \right) dx \qquad (4.7)$$

within $\mathcal{K}$. Here, the time-discrete term $\mathcal{M}$ has several possible forms, as already mentioned in the beginning of this chapter:

$$\mathcal{M}^O = \int_{\Omega \cap S_n(u)} \frac{|u - 2u_{n-1} + u_{n-2}|^2}{2h^2} dx,$$

$$S_n(u) := \{u > 0\} \cup \{u_{n-1} > 0\} \cup \{u_{n-2} > 0\},$$

$$\mathcal{M}^K = \int_\Omega \frac{|u - 2u_{n-1} + u_{n-2}|^2}{2h^2} dx + I(u), \quad I(u) := \begin{cases} 0 & \text{if } u \geq 0 \text{ a.e. in } \Omega, \\ \infty & \text{if otherwise} \end{cases},$$

$$\mathcal{M}^S = \int_\Omega \frac{u - 4u_{n-1} + 2u_{n-2}}{2h^2} u \chi_{\{u>0\}} dx. \qquad (4.8)$$

$\mathcal{M}^O$ is from [92], $\mathcal{M}^K$ was presented in [34], and $\mathcal{M}^S$ in [83].

The paper [34], which uses $\mathcal{M}^K$, deals with the one-dimensional problem $\Omega = (0, 1)$ with $F = 0$ in the form

$$\begin{cases} u_{tt} - u_{xx} \geq 0 & \text{in} \quad \Omega \times (0, T), \\ \text{spt}\{u_{tt} - u_{xx}\} \subset \{u = 0\}, \\ u(x, t) \geq 0 & \text{for} \quad \mathcal{L}^2\text{- a.e. } (x, t), \end{cases}$$

under suitable boundary and initial conditions, which is equivalent to

$$u_{tt} - u_{xx} + \partial I(u) \ni 0.$$

Unlike the Yoshida approximation approach adopted in [52, 53], the author applies the discrete Morse flow (4.7) with $\mathcal{M}^K$ and passes to the limit as $h \to 0+$ in the weak form of the corresponding Euler-Lagrange equation

$$\frac{u - 2u_{n-1} + u_{n-2}}{h^2} - u_{xx} + \partial I(u) \ni 0,$$

obtaining a weak solution to the original problem. This limit works well because a uniform in h estimate of $\|u_t^h\|_{L^\infty(0,T;L^2(\Omega))}$ and $\|u_x^h\|_{L^\infty(0,T;L^2(\Omega))}$ is derived in a similar fashion to Lemma 3.1.

4.2 Construction of Approximate Solutions

We now proceed with an in-depth explanation of the discrete Morse flow method (4.7) using the form $\mathcal{M}^S$ of the discrete term. We will deal with the volume-preserving case. The approach using $\mathcal{M}^O$ will be treated in the next chapter in conjunction with energy preservation.

Nevertheless, both methods take analogous steps to construct a weak solution in the sense of Definition 4.1. If the minimizer is continuous, we can choose support of test functions inside $\{u > 0\}$, and derive the desired weak equation in the domain $\{u > 0\}$. The required continuity of minimizers is achieved by making use of the elliptic regularity theory of [24, 38, 56].

4.2.1 Assumptions

Let Ω be a bounded connected domain in $\mathbb{R}^m$ with Lipschitz boundary. Let $T > 0$ be the final time and $V > 0$ a prescribed volume. Suppose that initial functions $u_0, v_0 \in H_0^1(\Omega)$ are given. Moreover, let the potential of outer force $F(x, t, u)$ satisfy the assumptions (F1)–(F3).

4.2.2 Statement of Semi-discrete Problem

Let $N \in \mathbb{N}$ be given, set $h = T/N$, and define $u_{-1} = u_0 - h v_0$. For each $n = 1, 2, \ldots, N$, find a minimizer u_n of the functional

$$J_n(u) = \int_\Omega \frac{u - 4u_{n-1} + 2u_{n-2}}{2h^2} u \chi_{\{u>0\}}\, dx + \frac{1}{2} \int_\Omega |\nabla u|^2\, dx + \int_\Omega F_n(u)\, dx \tag{4.9}$$

in the function set

$$\mathcal{K}_V = \left\{ u \in H_0^1(\Omega);\ \int_\Omega u \chi_{\{u>0\}}\, dx = V \right\}.$$

Notice that the set $\mathcal{K}_V$ contains a special form of the volume constraint. Here, F_n is the average of the function F over the n-th time subinterval:

$$F_n(x, u) = \frac{1}{h} \int_{(n-1)h}^{nh} F(t, x, u)\, dt.$$

We also define a time-discretization of the outer force by

$$f_n(x, u) = \frac{1}{h} \int_{(n-1)h}^{nh} f(t, x, u)dt. \tag{4.10}$$

4.2.3 Minimizers Are Nonnegative

If a minimizer of J_n exists, it can be assumed to be nonnegative almost everywhere in Ω. Indeed, if $u \in \mathcal{K}_V$ is a minimizer, setting $\tilde{u} = u\chi_{\{u>0\}}$, we can prove that

$$J_n(\tilde{u}) \leq J_n(u). \tag{4.11}$$

To see this, note that $\tilde{u}$ belongs to $\mathcal{K}_V$. In addition, the first term of $J_n(\tilde{u})$ on the right-hand side of (4.9) is equal to the first term of $J_n(u)$, the Dirichlet term for $\tilde{u}$ is less than or equal to that of u, and the same holds for the last outer force term thanks to assumption (F2). Hence (4.11) follows. We have used the fact that $\nabla \max\{u, 0\}(x) = 0$ a.e. in $\{x : u(x) \leq 0\}$ (see [25, Chapter 7.4]). To show strict inequality in (4.11) when u is not nonnegative a.e., would require finer analysis but it is enough for us to know that there is a nonnegative minimizer.

4.2.4 Existence of Minimizers

We state this result as a theorem.

Theorem 4.1 *There exists a nonnegative minimizer u_n of the functional J_n in $\mathcal{K}_V$.*

Proof For given $u_{n-1}, u_{n-2} \in \mathcal{K}_V$, we shall show the existence of a minimizer u_n. We prove lower semicontinuity of J_n, which automatically leads to existence. Since the minimizer is supposed to be nonnegative, it is sufficient to show the existence in $\mathcal{K}'_V = \{u \in \mathcal{K}_V; u \geq 0\}$, which is a closed convex set.

Since

$$\int_\Omega \frac{u - 4u_{n-1} + 2u_{n-2}}{2h^2} u\chi_{\{u>0\}}\, dx$$

$$= \int_\Omega \frac{(u - 2u_{n-1} + u_{n-2})^2 - (2u_{n-1} - u_{n-2})^2}{2h^2}\chi_{\{u>0\}}\, dx$$

$$\geq -\int_\Omega \frac{(2u_{n-1} - u_{n-2})^2}{2h^2}\, dx,$$

the boundedness of the functional from below is immediate when $F \equiv 0$. In the case of outer force being present, we utilize the assumption (F3) to get for any

$u \in H_0^1(\Omega)$,

$$\int_\Omega F_n(u)\, dx = \frac{1}{h} \int_\Omega \int_{(n-1)h}^{nh} F(x,t,u)\, dt\, dx$$

$$\geq -\frac{1}{h} \int_\Omega \int_{(n-1)h}^{nh} \left(C_f \, |u|^2 + \vartheta(x,t)\, |u| \right) dt\, dx$$

$$\geq -\int_\Omega \left(C_f \, |u|^2 + \sup_{t\in(0,T)} |\vartheta(x,t)|\, |u| \right) dx.$$

The last expression is for $C_f > 0$ greater than or equal to

$$-\left(C_f + \frac{C_f}{2} \right) \int_\Omega |u|^2\, dx - \frac{1}{2C_f} \int_\Omega \sup_{t\in(0,T)} \vartheta^2(x,t)\, dx,$$

and for $C_f = 0$ it is greater than or equal to

$$-\frac{3}{8C_P^2} \int_\Omega |u|^2\, dx - \frac{2C_P^2}{3} \int_\Omega \sup_{t\in(0,T)} \vartheta^2(x,t)\, dx.$$

In either case, by (F3) we find that

$$\int_\Omega F_n(u)\, dx \geq -\frac{3}{8C_P^2} \int_\Omega |u|^2\, dx - C\|\vartheta\|_{L^\infty(0,T;L^2(\Omega))}^2$$

$$\geq -\frac{3}{8} \int_\Omega |\nabla u|^2\, dx - C\|\vartheta\|_{L^\infty(0,T;L^2(\Omega))}^2$$

for some constant C. Consequently,

$$J_n(u) \geq -\int_\Omega \frac{(2u_{n-1} - u_{n-2})^2}{2h^2}\, dx + \frac{1}{8} \int_\Omega |\nabla u|^2\, dx - C\|\vartheta\|_{L^\infty(0,T;L^2(\Omega))}^2,$$

$$\tag{4.12}$$

which is bounded below by a fixed constant independent of u.

Hence, there exists a minimizing sequence $\{u^k\} \subset \mathcal{K}_V'$ of J_n:

$$\lim_{k\to\infty} J_n(u^k) = \inf_{u\in\mathcal{K}_V'} J_n(u).$$

In addition, thanks to (4.12) and Poincaré's inequality, u^k are uniformly bounded in $H^1(\Omega)$. Therefore, there is a subsequence, denoted by $\{u^k\}$ again, and a nonnegative limit function $u \in \mathcal{K}'_V$ such that

$$\nabla u^k \rightharpoonup \nabla u \qquad \text{weakly in } L^2(\Omega),$$

$$u^k \to u \qquad \text{almost everywhere in } \Omega,$$

$$u^k \to u \qquad \text{strongly in } L^2(\Omega).$$

Then, by the lower semicontinuity of Dirichlet integral and the assumptions (F1), (F3), we have

$$
J_n(u) = \int_\Omega \frac{u - 4u_{n-1} + 2u_{n-2}}{2h^2} u \, dx + \frac{1}{2} \int_\Omega |\nabla u|^2 \, dx + \int_\Omega F(u) \, dx
$$

$$
\leq \liminf_{k \to \infty} \left[\int_\Omega \frac{u^k - 4u_{n-1} + 2u_{n-2}}{2h^2} u^k \, dx + \frac{1}{2} \int_\Omega \left| \nabla u^k \right|^2 \, dx + \int_\Omega F(u^k) \, dx \right]
$$

$$
= \liminf_{k \to \infty} J_n(u^k),
$$

which means that $u \in \mathcal{K}_V$ is a minimizer, and this completes the proof. □

4.2.5 Energy Estimate

The basic energy estimate for discrete solutions is obtained in the same way as in problems without a free boundary, except for some technicalities related to the outer force term.

Proposition 4.1 *There is a constant C_E independent of n and h, such that minimizers u_n satisfy*

$$
\left\| \frac{u_n - u_{n-1}}{h} \right\|^2_{L^2(\Omega)} + \|\nabla u_n\|^2_{L^2(\Omega)} \leq C_E. \tag{4.13}
$$

Proof As the set $\mathcal{K}'_V = \{u \in \mathcal{K}_V; u \geq 0\}$ is convex, the function $(1 - \theta)u_n + \theta u_{n-1}$ belongs to $\mathcal{K}'_V$ for all $\theta \in [0, 1]$. By the minimality property we have

$$
0 \leq \frac{1}{\theta} \big(J_n((1 - \theta)u_n + \theta u_{n-1}) - J_n(u_n) \big).
$$

Passing to the limit as $\theta \to 0+$, we get

$$
0 \leq -\frac{1}{h^2} \int_\Omega (u_n - u_{n-1})(u_n - 2u_{n-1} + u_{n-2}) \, dx + \int_\Omega \nabla u_n \cdot \nabla(u_{n-1} - u_n) \, dx
$$

$$
+ \int_\Omega F'_n(u_n)(u_{n-1} - u_n) \, dx
$$

$$\leq \frac{1}{2h^2} \int_\Omega \left((u_{n-1} - u_{n-2})^2 - (u_n - u_{n-1})^2 \right) dx + \frac{1}{2} \int_\Omega \left(|\nabla u_{n-1}|^2 - |\nabla u_n|^2 \right) dx$$

$$+ \frac{h}{2} \int_\Omega (f_n(u_n))^2 \, dx + \frac{h}{2} \int_\Omega \left(\frac{u_n - u_{n-1}}{h} \right)^2 dx, \tag{4.14}$$

where f_n is defined by (4.10). The point here is that we can drop the characteristic function in the first term since our minimizers are nonnegative. The outer force term is estimated in the following way:

$$\int_\Omega (f_n(u_n))^2 \, dx = \int_\Omega \left(\frac{1}{h} \int_{(n-1)h}^{nh} f(x, t, u_n) \, dt \right)^2 dx$$

$$\leq \int_\Omega \left(\frac{1}{h} \int_{(n-1)h}^{nh} f^2(x, t, u_n) \, dt \right) dx$$

$$\leq \int_\Omega \left(\frac{1}{h} \int_{(n-1)h}^{nh} 2 \left(4C_f^2 u_n^2 + \vartheta^2(x, t) \right) dt \right) dx$$

$$\leq 8C_f^2 \|u_n\|_{L^2(\Omega)}^2 + 2|\Omega| \, \|\vartheta\|_{L^\infty(Q_T)}^2$$

$$\leq \|\nabla u_n\|_{L^2(\Omega)}^2 + 2|\Omega| \, \|\vartheta\|_{L^\infty(Q_T)}^2.$$

We have used assumption (F3) again. Thus, plugging this estimate into (4.14) and summing up over n from 1 to n, we arrive at

$$\left\| \frac{u_n - u_{n-1}}{h} \right\|_{L^2(\Omega)}^2 + \|\nabla u_n\|_{L^2(\Omega)}^2 \leq \|v_0\|_{L^2(\Omega)}^2 + \|\nabla u_0\|_{L^2(\Omega)}^2 + 2|\Omega| T \|\vartheta\|_{L^\infty(Q_T)}^2$$

$$+ h \sum_{k=1}^n \left(\left\| \frac{u_k - u_{k-1}}{h} \right\|_{L^2(\Omega)}^2 + \|\nabla u_n\|_{L^2(\Omega)}^2 \right).$$

The discrete Gronwall lemma implies

$$\left\| \frac{u_n - u_{n-1}}{h} \right\|_{L^2(\Omega)}^2 + \|\nabla u_n\|_{L^2(\Omega)}^2 \leq \left(\|v_0\|_{L^2(\Omega)}^2 + \|\nabla u_0\|_{L^2(\Omega)}^2 + 2|Q_T| \|\vartheta\|_{L^\infty(Q_T)}^2 \right) e^{2T}.$$

This is already the desired estimate. $\square$

4.2.6 Subsolution Property

We next prove an inequality that is satisfied by the discrete minimizers u_n. We consider this inequality important since it in a certain sense guarantees the positiveness of the linear functional corresponding to first variation of J_n. In elliptic theory, this fact allows us to identify the operator with a Radon measure concentrated on the free boundary.

Theorem 4.2 *A nonnegative minimizer u of J_n satisfies the following inequality for arbitrary nonnegative function $\zeta \in C_0^\infty(\Omega)$:*

$$\int_\Omega \left(\frac{u - 2u_{n-1} + u_{n-2}}{h^2} \chi_{\{u>0\}}\zeta + \nabla u \cdot \nabla \zeta + F_n'(u)\zeta \right) dx \tag{4.15}$$

$$- \frac{1}{V} \left(\int_\Omega \left(\frac{u - 2u_{n-1} + u_{n-2}}{h^2} u + |\nabla u|^2 + F_n'(u)u \right) dx \right) \left(\int_{\{u>0\}} \zeta \, dx \right) \leq 0.$$

Proof Without loss of generality, we can simplify the calculations by taking $V = 1$. For $\delta > 0$, let us set

$$\psi_\delta := \frac{u - \delta\zeta}{I_\delta} \chi_{\{u-\delta\zeta>0\}}, \qquad \text{where} \ \ I_\delta = \int_\Omega (u - \delta\zeta)\chi_{\{u-\delta\zeta>0\}} \, dx.$$

We have $I_\delta \leq 1$ by the volume condition $\int_\Omega u \, dx = V = 1$, while

$$I_\delta = \int_\Omega u\chi_{\{u-\delta\zeta>0\}} \, dx - \delta \int_\Omega \zeta\chi_{\{u-\delta\zeta>0\}} \, dx$$

$$= \int_\Omega u \, dx - \int_{\{u\leq\delta\zeta\}} u \, dx - \delta \int_{\{u>\delta\zeta\}} \zeta \, dx,$$

which yields

$$1 - I_\delta = \int_{\{u\leq\delta\zeta\}} u \, dx + \delta \int_{\{u>\delta\zeta\}} \zeta \, dx$$

$$= \delta \int_{\{u>0\}} \zeta \, dx + \int_{\{u-\delta\zeta\leq0\}} (u - \delta\zeta) \, dx - \int_{\{u=0\}} (u - \delta\zeta) \, dx$$

$$= \delta \int_{\{u>0\}} \zeta \, dx + \int_{\{0<u\leq\delta\zeta\}} (u - \delta\zeta) \, dx. \tag{4.16}$$

Noting that

$$\left| \int_{\{0<u\leq\delta\zeta\}} (u - \delta\zeta) \, dx \right| \leq 2\delta \int_{\{0<u\leq\delta\zeta\}} \zeta \, dx \leq 2\delta \|\zeta\|_{L^\infty(\Omega)} |\{0 < u \leq \delta\zeta\}| = o(\delta),$$

this implies

$$\lim_{\delta\to0} \frac{1 - I_\delta}{\delta} = \int_{\{u>0\}} \zeta \, dx. \tag{4.17}$$

In order to shorten the upcoming estimates, we write $Z := \int_{\{u>0\}} \zeta \, dx$.

As $\psi_\delta \in \mathcal{K}_V$, we have $J_n(\psi_\delta) - J_n(u) \geq 0$ by the minimizing property of u. Hence,

$$
\begin{aligned}
0 \leq{} & \int_{\{u>\delta\zeta\}} \frac{\frac{u-\delta\zeta}{I_\delta} - 4u_{n-1} + 2u_{n-2}}{2h^2} \frac{u-\delta\zeta}{I_\delta}\, dx - \int_\Omega \frac{u - 4u_{n-1} + 2u_{n-2}}{2h^2} u\, dx \\
& + \int_{\{u>\delta\zeta\}} \frac{|\nabla(u-\delta\zeta)|^2}{2I_\delta^2}\, dx - \int_\Omega \frac{1}{2}|\nabla u|^2\, dx \\
& + \int_\Omega \left[F_n\left(\frac{u-\delta\zeta}{I_\delta} \chi_{\{u-\delta\zeta>0\}} \right) - F_n(u) \right] dx \\
={} & \int_{\{u>\delta\zeta\}} \left[\frac{u-\delta\zeta - 4u_{n-1} + 2u_{n-2}}{2h^2}(u-\delta\zeta) - \frac{u - 4u_{n-1} + 2u_{n-2}}{2h^2} u \right] dx \\
& + \int_{\{u>\delta\zeta\}} \left[\frac{\frac{u-\delta\zeta}{2I_\delta} - 2u_{n-1} + u_{n-2}}{h^2} \frac{u-\delta\zeta}{I_\delta} - \frac{\frac{u-\delta\zeta}{2} - 2u_{n-1} + u_{n-2}}{h^2}(u-\delta\zeta) \right] dx \\
& - \int_{\{0<u\leq\delta\zeta\}} \frac{u - 4u_{n-1} + 2u_{n-2}}{2h^2} u\, dx \\
& + \frac{1}{2} \int_{\{u>\delta\zeta\}} \left[|\nabla(u-\delta\zeta)|^2 - |\nabla u|^2 \right] dx + \frac{1}{2}\int_{\{u>\delta\zeta\}} \left(\frac{1}{I_\delta^2} - 1 \right)|\nabla(u-\delta\zeta)|^2\, dx \\
& - \frac{1}{2}\int_{\{0<u\leq\delta\zeta\}} |\nabla u|^2\, dx + \int_\Omega f_n(\widetilde u)\left(\frac{u-\delta\zeta}{I_\delta}\chi_{\{u-\delta\zeta>0\}} - u \right) dx,
\end{aligned}
$$

where $\widetilde u(x)$ is a value between $u(x)$ and $\psi_\delta(x)$. This is further equal to

$$
\begin{aligned}
={} & -\int_{\{u>\delta\zeta\}} \delta \frac{u - 2u_{n-1} + u_{n-2}}{h^2} \zeta\, dx - \frac{\delta}{2h^2}\int_{\{u>\delta\zeta\}} \zeta(u-\delta\zeta)\, dx \\
& + \int_{\{u>\delta\zeta\}} \frac{(u^2 - 2\delta u\zeta + \delta^2\zeta^2)(1 - I_\delta^2) - (4u_{n-1} - 2u_{n-2})(u-\delta\zeta)(I_\delta - I_\delta^2)}{2h^2 I_\delta^2}\, dx \\
& - \int_{\{0<u\leq\delta\zeta\}} \frac{u - 4u_{n-1} + 2u_{n-2}}{2h^2} u\, dx - \frac{1}{2}\int_{\{0<u\leq\delta\zeta\}} |\nabla u|^2\, dx \\
& + \int_{\{u>\delta\zeta\}} \left[-\delta\nabla u \cdot \nabla\zeta + \frac{1}{2}\delta^2|\nabla\zeta|^2 \right] dx + \int_{\{u>\delta\zeta\}} \frac{1 - I_\delta^2}{2I_\delta^2}|\nabla(u-\delta\zeta)|^2\, dx \\
& + \int_\Omega f_n(\widetilde u)\left(\left(\frac{u(1 - I_\delta)}{I_\delta} - \frac{\delta\zeta}{I_\delta} \right)\chi_{\{u-\delta\zeta>0\}} - u\chi_{\{0<u\leq\delta\zeta\}} \right) dx \\
\leq{} & -\delta\int_{\{u>\delta\zeta\}} \frac{u - 2u_{n-1} + u_{n-2}}{h^2} \zeta\, dx \\
& + \int_{\{u>\delta\zeta\}} \frac{2\delta Zu - (4u_{n-1} - 2u_{n-2})(I_\delta - I_\delta^2)}{2h^2 I_\delta^2} u\, dx
\end{aligned}
$$

$$-\int_{\{0<u\leq\delta\zeta\}}\frac{u-4u_{n-1}+2u_{n-2}}{2h^2}u\,dx+\int_{\{u>\delta\zeta\}}\frac{\delta Z}{I_\delta^2}|\nabla u|^2\,dx$$

$$-\delta\int_{\{u>\delta\zeta\}}\nabla u\cdot\nabla\zeta\,dx+\int_{\{u>\delta\zeta\}}f_n(\tilde{u})\left(\frac{u(1-I_\delta)}{I_\delta}-\frac{\delta\zeta}{I_\delta}\right)dx+o(\delta).$$

Using the nonnegativity of u and u_{n-2}, we can estimate the third term as

$$-\int_{\{0<u\leq\delta\zeta\}}\frac{u-4u_{n-1}+2u_{n-2}}{2h^2}u\,dx\leq\frac{2}{h^2}\int_{\{0<u\leq\delta\zeta\}}uu_{n-1}\,dx$$

$$\leq\frac{2}{h^2}\int_{\{0<u\leq\delta\zeta\}}\delta\max\zeta\cdot u_{n-1}\,dx$$

$$\leq\frac{2\delta\max\zeta}{h^2}\|u_{n-1}\|_{L^2(\Omega)}\,|\{0<u\leq\delta\zeta\}|^{1/2}.$$

Moreover, since $u-\psi_\delta=o(1)$ and f_n is continuous in u and satisfies (F3), the limit as $\delta\to0$ in the last term involving f_n can be handled by Lebesgue's dominated convergence theorem.

Hence, dividing by δ, taking $\delta\to0+$ in the above and using (4.17), we obtain

$$0\leq-\int_{\{u>0\}}\frac{u-2u_{n-1}+u_{n-2}}{h^2}\zeta\,dx+\int_{\{u>0\}}\frac{u-2u_{n-1}+u_{n-2}}{h^2}uZ\,dx$$

$$+\lim_{\delta\to0+}\frac{2\max\zeta}{h^2}\|u_{n-1}\|_{L^2(\Omega)}\,|\{0<u\leq\delta\zeta\}|^{1/2}$$

$$-\int_\Omega\nabla u\cdot\nabla\zeta\,dx+\int_\Omega|\nabla u|^2Z\,dx-\int_\Omega f_n(u)\zeta\,dx+\int_\Omega f_n(u)uZ\,dx$$

$$=-\int_{\{u>0\}}\frac{u-2u_{n-1}+u_{n-2}}{h^2}\zeta\,dx+\int_{\{u>0\}}\frac{u-2u_{n-1}+u_{n-2}}{h^2}uZ\,dx$$

$$-\int_\Omega\nabla u\cdot\nabla\zeta\,dx+\int_\Omega|\nabla u|^2Z\,dx-\int_\Omega f_n(u)\zeta\,dx+\int_\Omega f_n(u)uZ\,dx.$$

This is the announced inequality.   □

4.2.7 L^∞-Boundedness

If we additionally assume that $u_0,v_0\in H^1(\Omega)\cap L^\infty(\Omega)$ then minimizers fulfill the bound $\|u_n\|_{L^\infty(\Omega)}\leq M<\infty$, where M may depend on h.

Without loss of generality, we can set $V=1$. The statement is proven by a standard elliptic technique of [38], but here we have to consider a test function belonging to $\mathcal{K}_V$, particularly of the form

$$\psi_\delta(u) = \frac{u - \delta(u-k)^+}{\int_\Omega (u - \delta(u-k)^+)\,dx}, \qquad \delta > 0,$$

where $(u - k)^+ = \max\{u - k, 0\}$ and $k \geq 1$. We use mathematical induction. For $n = 1$, the functions $u_{n-2} = u_0 - hv_0$ and $u_{n-1} = u_0$ are bounded, so we assume that we have the boundedness of u_k for $k = 1, \ldots, n-1$, and prove the boundedness of u_n. We calculate

$$\lim_{\delta \to 0+} \frac{J_n(\psi_\delta(u_n)) - J_n(u_n)}{\delta} \geq 0.$$

Minimizer u_n is nonnegative, and if $\delta < 1$ then $\psi_\delta(u_n)$ is also nonnegative. Thus we can omit the characteristic function appearing in the first term of the expressions for $J_n(u_n)$ and $J_n(\psi_\delta(u_n))$. This yields after taking δ to zero the relation

$$\int_\Omega \left(\frac{u_n - 2u_{n-1} + u_{n-2}}{h^2}(u_n - k)^+ + \nabla u_n \cdot \nabla(u_n - k)^+ + F_n'(u_n)(u_n - k)^+ \right) dx$$

(4.18)

$$\leq \left(\int_\Omega (u_n - k)^+ \, dx \right) \int_\Omega \left(\frac{u_n - 2u_{n-1} + u_{n-2}}{h^2} u_n + |\nabla u_n|^2 + F_n'(u_n)u_n \right) dx.$$

The last integral on the right-hand side of the above inequality can be estimated in the following way, where we use the energy estimate (4.13) and (F3):

$$\int_\Omega \left(\frac{u_n - 2u_{n-1} + u_{n-2}}{h^2} u_n + |\nabla u_n|^2 + F_n'(u_n)u_n \right) dx$$

(4.19)

$$\leq C + (2C_f + 1)\left(\|u_n\|_{L^2(\Omega)}^2 + \|\vartheta\|_{L^2(\Omega)}^2 \right)$$

$$\leq C.$$

Here C is a generic constant that depends on h and on the L^∞-bound of u_{n-1} and u_{n-2}. Let us set $A_k = \{x \in \Omega : u_n(x) > k\}$. Neglecting terms with appropriate sign and noting that $(u_n - k)^+ = 0$ on $\Omega \setminus A_k$, we have from (4.18)

$$\int_{A_k} |\nabla u_n|^2 \, dx$$

$$\leq \int_{A_k} -\frac{u_n - 2u_{n-1} + u_{n-2}}{h^2}(u_n - k)\,dx + \int_{A_k} (|f_n(u_n)| + C)(u_n - k)\,dx$$

$$\leq \frac{2}{h^2} \int_{A_k} u_{n-1}(u_n - k)\,dx + 2C_f \int_{A_k} u_n(u_n - k)\,dx + C \int_{A_k} (u_n - k)\,dx$$

$$\leq C_f \left(3 \int_{A_k} (u_n - k)^2 \, dx + k^2 |A_k| \right) + C \int_{A_k} (u_n - k) \, dx$$

$$\leq C \left(\int_{A_k} (u_n - k)^2 \, dx + k^2 |A_k| \right).$$

This inequality gives the estimate assumed in Theorem 2.5.1 from [38] to obtain the bound for u_n.

4.2.8 Hölder Continuity

Theorem 4.3 *For any compact subset $\widetilde{\Omega} \Subset \Omega$, there exists a positive constant α (depending on h) with $0 < \alpha < 1$, such that minimizers u_n belong to $C^{0,\alpha}(\widetilde{\Omega})$.*

Proof We shall show that minimizers belong to the space $\mathcal{B}_2(\Omega, M, \gamma, d)$ defined below. Results from Chapter 2.6 in [38] then ensure their Hölder continuity.

Definition 4.3 A function $u \in H^1(\Omega)$ belongs to the class $\mathcal{B}_2(\Omega, M, \gamma, d)$ if

(1) $M = \sup_{\Omega} |u| < \infty$,
(2) There is $\gamma > 0$ such that, for $w = \pm u$,

$$\int_{A_{k,r-\sigma r}} |\nabla w|^2 \, dx \leq \gamma \left[\frac{1}{(\sigma r)^2} \sup_{B_r} (w - k)^2 + 1 \right] |A_{k,r}|, \tag{4.20}$$

for all $\sigma \in (0, 1)$, $B_r \subset \Omega$ and k with $k \geq \max_{B_r} w - d$, and for any $d > 0$. Here $A_{k,r} = \{x \in B_r; \; w(x) > k\}$ and the symbol B_r means a ball of radius r.

Without loss of generality, we can set $V = 1$. Condition (1) from Definition 4.3 has already been proved above. Let us now derive estimate (4.20) for $w = +u$, using (4.15) for a suitable function ζ. Choose r and $s = r - \sigma r$, where $\sigma \in (0, 1)$, to be arbitrary numbers with $0 < s < r$. Take $\zeta = \eta^2 \max\{u - k, 0\}$, where η is a function with support in B_r satisfying $0 \leq \eta \leq 1$, $\eta = 1$ in B_s (concentric with B_r) and $|\nabla \eta| \leq 2/(r - s)$ in $B_r \setminus B_s$. Recalling the estimate of the multiplier (4.19) and neglecting positive terms in (4.15), we obtain the estimate

$$0 \leq \int_{A_{k,r}} \left(\frac{2u_{n-1}}{h^2} \chi_{\{u > 0\}} \zeta - \nabla u_n \cdot \nabla \zeta \right) dx + C \int_{\{u_n > 0\}} \zeta \, dx$$

$$\leq C |A_{k,r}| - \int_{\Omega} \nabla u_n \cdot \nabla \zeta \, dx, \tag{4.21}$$

with the constant C depending only on h, M and $|\Omega|$. The gradient term is estimated using Young's inequality:

$$
\begin{aligned}
&- \int_\Omega \nabla u_n \cdot \nabla \zeta \, dx \\
&= - \int_{A_{k,r}} |\nabla u_n|^2 \eta^2 \, dx - \int_{A_{k,r}} \nabla u_n \cdot \nabla \eta \, 2\eta (u_n - k) \, dx \\
&\le - \int_{A_{k,r}} |\nabla u_n|^2 \eta^2 \, dx + \frac{1}{2} \int_{A_{k,r}} |\nabla u_n|^2 \eta^2 \, dx + 2 \int_{A_{k,r}} |\nabla \eta|^2 (u_n - k)^2 \, dx \\
&\le - \frac{1}{2} \int_{A_{k,s}} |\nabla u_n|^2 \, dx + \frac{8}{(\sigma r)^2} \sup_{B_r} (u_n - k)^2 |A_{k,r}|.
\end{aligned}
$$

The above estimate together with (4.21) readily gives the desired result (4.20).

Let us turn to the case of $-u$. We set $w = -u$ and note that w minimizes the following functional:

$$
J_n(w) := \int_\Omega \left(\frac{w + 4u_{n-1} - 2u_{n-2}}{2h^2} w \chi_{\{w<0\}} + \frac{1}{2} |\nabla w|^2 + F_n(-w) \right) dx
\tag{4.22}
$$

among functions $w \in H_0^1(\Omega)$ satisfying $\int_\Omega w \chi_{\{w<0\}} = -V$. We introduce the function φ by the formula

$$
\varphi := -\frac{w - \zeta}{\int_\Omega (w - \zeta) \, dx},
\tag{4.23}
$$

where $\zeta = \eta \max\{w - k, 0\}$, the real number k is negative, and η is a smooth cut-off function chosen in the same way as above. For the sake of simplicity, let us write

$$
U := \frac{1}{\int_\Omega (w - \zeta) \, dx} = -\frac{1}{1 + \int_\Omega \zeta \, dx} \in [-1, 0).
$$

Since $w - \zeta$ is nonpositive, we see that φ belongs to $\mathcal{K}_{-V} := \{v \in H_0^1(\Omega); \int_\Omega v \chi_{\{v<0\}} = -V\}$. Thus $0 \le J_n(\varphi) - J_n(w)$. Using the nonpositivity of φ and w, the characteristic functions can be omitted, yielding

$$
\begin{aligned}
0 &\le \frac{1}{2h^2} \int_\Omega [(U(-w + \zeta) + 4u_{n-1} - 2u_{n-2})U(-w + \zeta) - (w + 4u_{n-1} - 2u_{n-2})w] \, dx \\
&\quad + \frac{U^2}{2} \int_\Omega |\nabla(-w + \zeta)|^2 \, dx - \frac{1}{2} \int_\Omega |\nabla w|^2 \, dx + \int_\Omega \left(F_n(U(w - \zeta)) - F_n(-w) \right) dx \\
&\le \frac{1}{2h^2} \int_\Omega [(1 + U)(-4u_{n-1} + 2u_{n-2})w + (U^2 - 1)w^2
\end{aligned}
$$

$$+ U\zeta\,(4u_{n-1} - 2u_{n-2} - 2Uw + U\zeta)\big]\,dx$$

$$+ \frac{U^2}{2}\int_\Omega |\nabla(-w+\zeta)|^2\,dx - \frac{1}{2}\int_\Omega |\nabla w|^2\,dx + \int_\Omega f_n(\widetilde{w})\big(-U\zeta + (1+U)w\big)\,dx$$

$$\leq C\int_\Omega \zeta\,dx + \frac{U^2}{2}\int_\Omega |\nabla(-w+\zeta)|^2\,dx - \frac{1}{2}\int_\Omega |\nabla w|^2\,dx. \tag{4.24}$$

Here we used the fact that $U^2 - 1 \leq 0$, that $1 + U \leq \int_\Omega \zeta\,dx$ and that f satisfies (F3). We estimate the terms containing gradient on the right-hand side of the above inequality. Noting that $U^2 \leq 1$ we can write

$$\frac{U^2}{2}\int_\Omega |\nabla(-w+\zeta)|^2\,dx - \frac{1}{2}\int_\Omega |\nabla w|^2\,dx$$

$$\leq \frac{U^2}{2}\int_{A_{k,r}} |(\eta-1)\nabla w + (w-k)\nabla\eta|^2\,dx + \frac{U^2}{2}\int_{\Omega\backslash A_{k,r}} |\nabla w|^2\,dx - \frac{1}{2}\int_\Omega |\nabla w|^2\,dx$$

$$\leq U^2 \int_{A_{k,r}} \left(|\nabla w|^2(1-\eta)^2 + |\nabla\eta|^2(w-k)^2\right)\,dx - \frac{1}{2}\int_{A_{k,r}} |\nabla w|^2\,dx$$

$$\leq U^2 \int_{A_{k,r}} |\nabla w|^2\,dx + U^2 \int_{A_{k,r}} |\nabla\eta|^2(w-k)^2\,dx - \left(\frac{1}{2}+U^2\right)\int_{A_{k,s}} |\nabla w|^2\,dx.$$

We then have by (4.24)

$$\int_{A_{k,s}} |\nabla w|^2\,dx \leq 2C|A_{k,r}| + \theta \int_{A_{k,r}} |\nabla w|^2\,dx + \frac{4}{(r-s)^2}\int_{A_{k,r}} (w-k)^2\,dx,$$

where the constant C again depends only on M, $|\Omega|$, ε and h, and $\theta = U^2/(1/2 + U^2) < 1$. Applying Lemma V.3.1 from [24], we obtain the desired estimate (4.20). $\qquad\square$

4.2.9   Euler–Lagrange Equation

Based on the preceding continuity result, we can choose the support of test functions in the set $\{u_n > 0\}$. Then the first variation of $J_n(u)$ yields the following weak formulae:

$$\int_\Omega \left(\frac{u_n - 2u_{n-1} + u_{n-2}}{h^2}\phi + \nabla u_n \cdot \nabla\phi + f_n(u_n)\phi\right)\,dx = \int_\Omega \lambda_n\phi\,dx,$$

$$\phi \in C_0^\infty(\Omega \cap \{u_n > 0\}),$$

$$\int_\Omega \nabla u_n \cdot \nabla\phi\,dx = 0, \quad \phi \in C_0^\infty(\Omega \cap \{u_n \leq 0\}^\circ).$$

Here,

$$\lambda_n = \frac{1}{V} \int_\Omega \left(\frac{u_n - 2u_{n-1} + u_{n-2}}{h^2} u_n + |\nabla u_n|^2 + f_n(u_n)u_n \right) dx$$

is the Lagrange multiplier coming from the volume constraint. From the second identity, we can conclude that $u_n \equiv 0$ outside the set $\{u_n > 0\}$.

4.2.10   Approximate Weak Solution

In order to define an approximation to a weak solution, we first introduce interpolated functions $\bar{u}^h$, u^h, $\bar{\lambda}^h$ and $\bar{f}^h$ on $\Omega \times (0, T)$ by

$$\bar{u}^h(x, t) = u_n(x),$$

$$u^h(x, t) = \frac{t - (n-1)h}{h} u_n(x) + \frac{nh - t}{h} u_{n-1}(x),$$

$$\bar{\lambda}^h(t) = \lambda_n,$$

$$\bar{f}^h(x, t, \bar{u}^h) = f_n(x, u_n),$$

where $(x, t) \in \Omega \times ((n-1)h, nh]$, $n = 0, 1, \ldots, N$ but we use only $n = 1, \ldots, N$ for $\bar{\lambda}^h$.

Definition 4.4 Let $\{u_n\} \subset \mathcal{K}_V$ and let $\bar{u}^h$ and u^h be defined as above. If the identities

$$\int_0^T \int_\Omega \left(\frac{u_t^h(t) - u_t^h(t-h)}{h} \phi + \nabla \bar{u}^h \cdot \nabla \phi + \bar{f}^h(\bar{u}^h)\phi \right) dx\, dt = \int_0^T \int_\Omega \bar{\lambda}^h \phi\, dx\, dt,$$

$$\forall \phi \in C_0^\infty(\Omega \times [0, T) \cap \{u^h > 0\}), \tag{4.25}$$

$$u^h \equiv 0 \quad \text{in} \quad Q_T \setminus \{u^h > 0\}, \tag{4.26}$$

and the initial conditions $u^h(0) = u_0$, $u^h(-h) = u_0 - hv_0$ are satisfied, then we call $\bar{u}^h$ and u^h approximate solutions.

The ultimate goal is to take $h \to 0+$ in (4.25) and recover the definition of a weak solution (4.6). For that, uniform estimates are needed and should be provided by the energy estimate (4.13), which rewritten in terms of $\bar{u}^h$ and u^h becomes

$$\|u_t^h(t)\|_{L^2(\Omega)}^2 + \|\nabla \bar{u}^h(t)\|_{L^2(\Omega)}^2 \leq C \qquad \text{for a.e. } t \in (0, T). \tag{4.27}$$

Unfortunately, the convergent subsequence that can be deduced from this estimate is not sufficient to pass to the limit in (4.25). Indeed, (4.27) yields a subsequence such that $u_t^h, \nabla \overline{u}_t^h$ converge weakly* in $L^\infty(0, T; L^2(\Omega))$ and $u^h, \overline{u}^h$ converge strongly in $L^2(Q_T)$, but this does not allow us to conclude that $\operatorname{spt} \phi \subset \{u > 0\}$ implies $\operatorname{spt} \phi \subset \{u^h > 0\}$ for small h. This property can be proved if $\dim(\Omega) = 1$ by extracting a uniformly convergent subsequence based on the Arzelà–Ascoli theorem. Accordingly, the limit $h \to 0+$ is possible and leads to a weak solution in accordance with Definition 4.2.

Still, we have succeeded in the construction of a quite reasonable approximate solution to the original hyperbolic free boundary problem. In addition, these approximate solutions converge to a function $u \in H^1(Q_T)$, which is expected to be the desired weak solution. However, this final limit step remains an open question, except for the one-dimensional problem.

4.3 Existence of Weak Solution in One Dimension

We present the limit process for the one-dimensional problem in this section. For simplicity, we shall do so only for the problem without volume constraint, i.e., we shall set the discrete and interpolated Lagrange multipliers $\lambda_n, \overline{\lambda}^h$ to zero in the above derivation, and prove the existence of weak solution in the sense of Definition 4.1. We refer to [26, 83] for the analysis of the volume-constrained problem.

4.3.1 Convergent Subsequence

First we extract a subsequence of approximate solutions converging in a sufficiently strong topology.

Lemma 4.1 *Let Ω be an open interval in $\mathbb{R}$. Then there is a decreasing sequence $\{h_j\}_{j=1}^\infty$ with limit zero (we write only $h \to 0+$) and a function $u \in H^1(0, T; L^2(\Omega)) \cap L^\infty(0, T; H_0^1(\Omega))$ such that*

$$u_t^h \rightharpoonup u_t \qquad \text{weakly* in } L^\infty(0, T; L^2(\Omega)), \tag{4.28}$$

$$\nabla \overline{u}^h \rightharpoonup \nabla u \qquad \text{weakly* in } L^\infty(0, T; L^2(\Omega)), \tag{4.29}$$

$$u^h \rightrightarrows u \qquad \text{uniformly in } Q_T, \tag{4.30}$$

The cluster function u is continuous and nonnegative in Q_T and satisfies the initial condition $u(0, x) = u_0(x)$ in Ω.

Proof The first two statements follow from the estimate (4.27). We shall prove uniform equicontinuity of u^h with respect to h. First, we see by (4.27) that

$$\|u^h(s) - u^h(t)\|^2_{L^2(\Omega)} = \int_\Omega \left(\int_t^s u_t^h(\tau)\, d\tau \right)^2 dx \le C(s - t),$$

with C independent of h. Further, using the inequality

$$\|g\|_{L^\infty(\Omega)} \le C\|g\|^{1/2}_{L^2(\Omega)} \left\| \frac{dg}{dx} \right\|^{1/2}_{L^2(\Omega)},$$

which holds in one dimension for any $g \in H_0^1(\Omega)$, we find again by (4.27) that

$$\|u^h(s) - u^h(t)\|_{L^\infty(\Omega)} \le C\|u^h(s) - u^h(t)\|^{1/2}_{L^2(\Omega)} \|u_x^h(s) - u_x^h(t)\|^{1/2}_{L^2(\Omega)} \le C|s - t|^{1/4}.$$

With these estimates at hand, the proof of equicontinuity is immediate:

$$|u^h(y, s) - u^h(x, t)| \le |u^h(y, s) - u^h(x, s)| + |u^h(x, s) - u^h(x, t)|$$

$$= \left| \int_x^y u_x^h(\xi, s)\, d\xi \right| + |u^h(x, s) - u^h(x, t)|$$

$$\le |y - x|^{1/2} \left(\int_\Omega |u_x^h(\xi, s)|^2\, d\xi \right)^{1/2} + C|s - t|^{1/4}$$

$$\le C\big(|y - x|^{1/2} + |s - t|^{1/4}\big).$$

Moreover, from this estimate we get the uniform boundedness in $L^\infty(Q_T)$ by setting $s = 0$ and $y = 0$. Therefore, by Arzelà–Ascoli theorem there is a subsequence $\{u^h\}$ converging to u uniformly in Q_T. $\qquad\square$

4.3.2 Limit Process

Now, we take h to zero in (4.25). As we wish to reach Definition 4.1, we fix an arbitrary $\phi \in C_0^\infty(\Omega \times [0, T) \cap \{u > 0\})$. Since u is continuous on Q_T, there is a constant $\eta > 0$ such that $u \ge \eta$ on spt ϕ. The subsequence $\{u^h\}$ from Lemma 4.1 converges to u uniformly, granting a positive h_0 so that

$$\max_{(x,t)\in Q_T} |u^h(x, t) - u(x, t)| \le \frac{\eta}{2} \qquad \forall h < h_0.$$

From this fact, we have $u^h \geq u - |u^h - u| \geq \eta/2$ on spt ϕ for all $h \in (0, h_0)$. Noting that $\bar{u}^h(x, t) = u^h(x, nh)$ $\forall t \in ((n - 1)h, nh]$, we have, in addition,

$$\bar{u}^h \geq \frac{\eta}{2} > 0 \quad \text{on spt } \phi \qquad \forall h \in (0, h_0).$$

This means that the relation

$$\int_0^T \int_\Omega \left(\frac{u_t^h(t) - u_t^h(t - h)}{h} \phi + \nabla \bar{u}^h \cdot \nabla \phi + \overline{f}^h(\bar{u}^h)\phi \right) dx \, dt = 0 \qquad (4.31)$$

holds for our test function ϕ, if $h < h_0$. Because of the outer force term, we need to apply a density argument and approximate the test function ϕ by a sequence of functions $\overline{\phi}^h$ which are piecewise constant on corresponding partitions of $(0, T)$ and which converge strongly to ϕ in $L^2\left(0, T; H^1(\Omega)\right)$. In particular, let us set

$$\overline{\phi}^h(x, t) = \frac{1}{h} \int_{(n-1)h}^{nh} \phi(x, \tau) \, d\tau$$

for $t \in ((n - 1)h, nh]$, $n = 1, \ldots, N$. Then we have

$$\int_0^T \int_\Omega \overline{f}^h(\bar{u}^h)\overline{\phi}^h \, dx \, dt = \int_0^T \int_\Omega f(\bar{u}^h)\overline{\phi}^h \, dx \, dt,$$

and since $(u_t^h(t) - u_t^h(t - h))/h$ is piecewise constant on partitions of the interval $(0, T)$, we also have

$$\int_0^T \int_\Omega \frac{u_t^h(t) - u_t^h(t - h)}{h} \overline{\phi}^h \, dx \, dt = \int_0^T \int_\Omega \frac{u_t^h(t) - u_t^h(t - h)}{h} \phi \, dx \, dt,$$

so that (4.31) becomes

$$\int_0^T \int_\Omega \left(\frac{u_t^h(t) - u_t^h(t - h)}{h} \phi + \nabla \bar{u}^h \cdot \nabla \overline{\phi}^h + f(\bar{u}^h)\overline{\phi}^h \right) dx \, dt = 0. \qquad (4.32)$$

Now we can pass to the limit as $h \to 0+$ in the term $\int_0^T \int_\Omega f(\bar{u}^h)\overline{\phi}^h \, dx \, dt$ because f is continuous in u, satisfies the bound (F3) necessary for the application of dominated convergence theorem, together with the boundedness of $\bar{u}^h$ and the pointwise convergence $\bar{u}^h \to u$.

The limit in the gradient term is calculated in the standard way. Namely,

$$\int_0^T \int_\Omega \nabla \bar{u}^h \cdot \nabla \overline{\phi}^h \, dx \, dt \to \int_0^T \int_\Omega \nabla u \cdot \nabla \phi \, dx \, dt \qquad \text{as } h \to 0+ \qquad (4.33)$$

follows immediately from (4.29) and the strong convergence of $\overline{\phi}^h$ in $L^2\left(0, T; H^1(\Omega)\right)$. Moreover, we have

$$\int_0^T \int_\Omega \frac{u_t^h(t) - u_t^h(t-h)}{h} \phi \, dx \, dt$$

$$= \int_0^T \int_\Omega \frac{u_t^h(t)}{h} \phi(t) \, dx \, dt - \int_{-h}^{T-h} \int_\Omega \frac{u_t^h(t)}{h} \phi(t+h) \, dx \, dt$$

$$= -\int_0^T \int_\Omega u_t^h(t) \frac{\phi(t+h) - \phi(t)}{h} \, dx \, dt - \int_{-h}^0 \int_\Omega \frac{u_t^h(t)}{h} \phi(t+h) \, dx \, dt$$

$$+ \int_{T-h}^T \int_\Omega \frac{u_t^h(t)}{h} \phi(t+h) \, dx \, dt$$

$$\to -\int_0^T \int_\Omega u_t \phi_t \, dx \, dt - \int_\Omega v_0 \phi(0) \, dx \qquad \text{as } h \to 0+. \qquad (4.34)$$

We arrive at

$$\int_0^T \int_\Omega (-u_t \phi_t + \nabla u \cdot \nabla \phi + f(u)\phi) \, dx \, dt - \int_\Omega v_0 \phi(0) \, dx = 0$$

for all $\phi \in C_0^\infty(\Omega \times [0, T) \cap \{u > 0\})$, which is formula (4.5) from Definition 4.1, while the conditions $u(x, 0) = u_0(x)$ and $u \equiv 0$ outside $\{u > 0\}$ are obviously fulfilled by construction.

Chapter 5
Energy-Preserving Discrete Morse Flow

In this chapter we focus on the energy-preserving scheme which was already touched upon in Sect. 3.2. What is different here is that we are dealing with a free boundary problem, while in Sect. 3.2, we solved a linear wave equation. In addition, we use this opportunity to describe the discrete Morse flow approach using the time-discretized term $\mathcal{M}^O$ from (4.8). To present these two ideas in one section should not cause any trouble because the energy-preserving modification is done in the gradient term of the functional, while the free boundary is handled mainly in the time-discrete term, and thus the ideas are clearly separated. Since our purpose is to focus on the conservation of energy, we will omit the outer force term and any global constraints from our considerations.

5.1 Modified Discrete Morse Flow Scheme

We are going to solve the same free boundary problem as in Chap. 4, i.e., we are searching for stationary points of the functional (4.2) in the set $\mathcal{K}^T$ from (4.4), except that we set $F \equiv 0$.

To define approximate solutions at discrete time instances, we take the initial conditions $u_0, v_0 \in H_0^1(\Omega)$ and define $u_{-1} = u_0 - hv_0$, where h is the time step obtained, as always, by dividing the time interval $(0, T)$ into N subintervals of equal length. Then for any integer $n = 1, \ldots, N$, we introduce the following functional:

$$J_n^{CN}(u) = \int_{\Omega \cap S_n(u)} \frac{|u - 2u_{n-1} + u_{n-2}|^2}{2h^2} \, dx + \frac{1}{4} \int_{\Omega} |\nabla u + \nabla u_{n-2}|^2 \, dx, \quad (5.1)$$

where $S_n(u) := \{u > 0\} \cup \{u_{n-1} > 0\} \cup \{u_{n-2} > 0\}$. We determine a sequence of functions $\{u_n\}$ iteratively by defining $\widetilde{u}_n$ as a minimizer of J_n^{CN} in $\mathcal{K} = H_0^1(\Omega)$, and setting $u_n := \max\{\widetilde{u}_n, 0\}$.

© The Author(s), under exclusive license to Springer Nature Singapore Pte Ltd. 2022
S. Omata et al., *Variational Approach to Hyperbolic Free Boundary Problems*,
SpringerBriefs in Mathematics, https://doi.org/10.1007/978-981-19-6731-3_5

We now study the existence and regularity of minimizers, which guarantee the possibility of applying the first variation formula to J_n^{CN} and thereby constructing approximate solutions.

Theorem 5.1 *There exists a minimizer $\widetilde{u}_n \in \mathcal{K}$ of the functional J_n^{CN}.*

Proof Given u_{n-1}, u_{n-2}, we show the existence of $\widetilde{u}_n$. Since the infimum of J_n^{CN} is nonnegative, we have only to show the lower semicontinuity of J_n^{CN}. Take any minimizing sequence $\{u^j\} \subset \mathcal{K}$ such that $J_n^{CN}(u^j) \to \inf_{u \in \mathcal{K}} J_n^{CN}(u)$ as $j \to \infty$. Since the sequence $\{u^j\}$ is bounded in $H^1(\Omega)$, there exist $\widetilde{u} \in H_0^1(\Omega)$ and $\gamma \in L^p(\Omega)$ for any $p \in [1, \infty)$ such that, up to extracting a subsequence,

$$u^j \to \widetilde{u} \qquad \text{strongly in } L^2(\Omega),$$

$$\nabla u^j \rightharpoonup \nabla \widetilde{u} \qquad \text{weakly in } L^2(\Omega), \tag{5.2}$$

$$\chi_{S_n(u^j)} \rightharpoonup \gamma \qquad \text{weakly in } L^p(\Omega).$$

Since $0 \leq \gamma \leq 1$ a.e. in Ω, and $\gamma = 1$ a.e. in $\{\widetilde{u} > 0\}$ (and thus also a.e. in $S_n(\widetilde{u})$), we have

$$J_n^{CN}(\widetilde{u}) = \int_\Omega \frac{|\widetilde{u} - 2u_{n-1} + u_{n-2}|^2}{2h^2} \chi_{S_n(\widetilde{u})} \, dx + \frac{1}{4} \int_\Omega |\nabla \widetilde{u} + \nabla u_{n-2}|^2 \, dx$$

$$\leq \int_\Omega \frac{|\widetilde{u} - 2u_{n-1} + u_{n-2}|^2}{2h^2} \gamma \, dx + \frac{1}{4} \int_\Omega |\nabla \widetilde{u} + \nabla u_{n-2}|^2 \, dx$$

$$\leq \liminf_{j \to \infty} J_n^{CN}(u^j),$$

where the second inequality follows from (5.2). $\square$

5.2 Evolution of Energy

Next, we show that minimizers of J_n^{CN} enjoy the energy-preserving property, at least in the regime where free boundary is not present. In the general regime with free boundary, we obtain an energy inequality.

Theorem 5.2 *The discrete energy*

$$E_n := \left\| \frac{u_n - u_{n-1}}{h} \right\|_{L^2(\Omega)}^2 + \frac{1}{2} \left(\|\nabla u_n\|_{L^2(\Omega)}^2 + \|\nabla u_{n-1}\|_{L^2(\Omega)}^2 \right), \tag{5.3}$$

does not increase in time in the following sense:

$$E_0 + E_1 \geq E_1 + E_2 \geq E_2 + E_3 \geq \cdots \geq E_{N-1} + E_N. \tag{5.4}$$

If, in addition, we suppose that $\Omega = \mathcal{S}_k(u_k) = \mathcal{S}_{k+1}(u_{k+1}) = \cdots = \mathcal{S}_{k+\ell}(u_{k+\ell})$ *up to a set of measure zero, for some integers* $k < k + \ell \leq N$, *then minimizers* $u_k, \ldots, u_{k+\ell}$ *conserve the energy in the sense that*

$$E_{k-1} = E_k = \cdots = E_{k+\ell}. \tag{5.5}$$

Proof Let us first prove the second statement (5.5). For n such that $k \leq n \leq k + \ell$, the function $(1 - \lambda)u_n + \lambda u_{n-2} = u_n + \lambda(u_{n-2} - u_n)$ is admissible for every $\lambda \in [0, 1]$. Since $\mathcal{S}_n(u_n) = \Omega$, the characteristic function in J_n^{CN} can be ignored, which justifies

$$\frac{d}{d\lambda} J_n^{CN}(u_n + \lambda(u_{n-2} - u_n))\bigg|_{\lambda=0} = 0.$$

Computing this derivative, we have

$$0 = \int_\Omega \left[\frac{(u_{n-2} - u_n)(u_n - 2u_{n-1} + u_{n-2})}{h^2} + \frac{1}{2}\nabla(u_{n-2} - u_n) \cdot \nabla(u_n + u_{n-2}) \right] dx$$

$$= \int_\Omega \left[\frac{(u_{n-1} - u_{n-2})^2 - (u_n - u_{n-1})^2}{h^2} + \frac{1}{2}|\nabla u_{n-2}|^2 - \frac{1}{2}|\nabla u_n|^2 \right] dx.$$

Summing over $n = k, \ldots, k + m$, where $1 \leq m \leq \ell$, we arrive at

$$\int_\Omega \bigg[\frac{1}{h^2}(u_{k-1} - u_{k-2})^2 - \frac{1}{h^2}(u_{k+m} - u_{k+m-1})^2$$

$$+ \frac{1}{2}|\nabla u_{k-2}|^2 + \frac{1}{2}|\nabla u_{k-1}|^2 - \frac{1}{2}|\nabla u_{k+m-1}|^2 - \frac{1}{2}|\nabla u_{k+m}|^2 \bigg] dx = 0,$$

which means

$$E_{k-1} = \int_\Omega \left[\frac{1}{h^2}(u_{k-1} - u_{k-2})^2 + \frac{1}{2}|\nabla u_{k-1}|^2 + \frac{1}{2}|\nabla u_{k-2}|^2 \right] dx$$

$$= \int_\Omega \left[\frac{1}{h^2}(u_{k+m} - u_{k+m-1})^2 + \frac{1}{2}|\nabla u_{k+m}|^2 + \frac{1}{2}|\nabla u_{k+m-1}|^2 \right] dx$$

$$= E_{k+m}.$$

In an analogous way, we get the equality of E_k and E_{k+m} for any $2 \leq m \leq \ell$. Since m was an arbitrary integer less than or equal to ℓ, the proof of (5.5) is complete.

Turning now to (5.4), we see that the function $(1-\lambda)\tilde{u}_n + \lambda u_{n-2} = \tilde{u}_n + \lambda(u_{n-2} - \tilde{u}_n)$ is admissible for any $\lambda \in [0, 1]$. Recall that $\tilde{u}_n$ is the symbol for a minimizer of J_n^{CN}. Hence, by the minimality property, we have $J_n^{CN}(\tilde{u}_n) \le J_n^{CN}(\tilde{u}_n + \lambda(u_{n-2} - \tilde{u}_n))$, and thus,

$$\lim_{\lambda \to 0+} \frac{1}{\lambda} \left(J_n^{CN}(\tilde{u}_n + \lambda(u_{n-2} - \tilde{u}_n)) - J_n^{CN}(\tilde{u}_n) \right) \ge 0. \tag{5.6}$$

Let A_n denote the set $\Omega \cap (\{\tilde{u}_n > 0\} \cup \{u_{n-1} > 0\} \cup \{u_{n-2} > 0\})$. We investigate the behavior of the individual terms in (5.6). For the gradient term we get

$$\lim_{\lambda \to 0+} \frac{1}{4\lambda} \left(|\nabla(\tilde{u}_n + \lambda(u_{n-2} - \tilde{u}_n) + u_{n-2})|^2 - |\nabla(\tilde{u}_n + u_{n-2})|^2 \right)$$

$$= \frac{1}{2} \nabla(\tilde{u}_n + u_{n-2}) \cdot \nabla(u_{n-2} - \tilde{u}_n)\, dx$$

$$= \frac{1}{2}|\nabla u_{n-2}|^2 - \frac{1}{2}|\nabla \tilde{u}_n|^2$$

$$\le \frac{1}{2}|\nabla u_{n-2}|^2 - \frac{1}{2}|\nabla u_n|^2. \tag{5.7}$$

For the time-discretized term, taking into account that the set

$$B_n(\lambda) := \{\tilde{u}_n + \lambda(u_{n-2} - \tilde{u}_n) > 0\} \cup \{u_{n-1} > 0\} \cup \{u_{n-2} > 0\}$$

is contained in the set A_n, we find that

$$\int_\Omega \left(|\tilde{u}_n + \lambda(u_{n-2} - \tilde{u}_n) - 2u_{n-1} + u_{n-2}|^2 \chi_{B_n(\lambda)} - |\tilde{u}_n - 2u_{n-1} + u_{n-2}|^2 \chi_{A_n} \right) dx$$

$$\le \int_\Omega \left(|\tilde{u}_n + \lambda(u_{n-2} - \tilde{u}_n) - 2u_{n-1} + u_{n-2}|^2 - |\tilde{u}_n - 2u_{n-1} + u_{n-2}|^2 \right) \chi_{A_n}\, dx.$$

Then we have

$$\frac{1}{2h^2\lambda} \int_\Omega \left(|\tilde{u}_n + \lambda(u_{n-2} - \tilde{u}_n) - 2u_{n-1} + u_{n-2}|^2 - |\tilde{u}_n - 2u_{n-1} + u_{n-2}|^2 \right) \chi_{A_n}\, dx$$

$$= \frac{1}{2h^2} \int_{A_n} (u_{n-2} - \tilde{u}_n)(2\tilde{u}_n + \lambda(u_{n-2} - \tilde{u}_n) - 4u_{n-1} + 2u_{n-2})\, dx$$

$$\xrightarrow[\lambda \to 0+]{} \frac{1}{h^2} \int_{A_n} (u_{n-2} - \tilde{u}_n)(\tilde{u}_n - 2u_{n-1} + u_{n-2})\, dx$$

$$= \frac{1}{h^2} \int_{A_n} \left[(u_{n-1} - u_{n-2})^2 - (u_{n-1} - \tilde{u}_n)^2 \right] dx. \tag{5.8}$$

Now, $\int_{A_n} (u_{n-1} - u_{n-2})^2 \, dx \leq \int_\Omega (u_{n-1} - u_{n-2})^2 \, dx$ since the integrand is nonnegative. Moreover, $u_n = \max\{\tilde{u}_n, 0\}$ and $u_{n-1} \geq 0$ imply $(u_{n-1} - \tilde{u}_n)^2 \geq (u_{n-1} - u_n)^2$. Therefore,

$$-\int_{A_n} (u_{n-1} - \tilde{u}_n)^2 \, dx \leq -\int_{A_n} (u_{n-1} - u_n)^2 \, dx.$$

Noting that, outside of A_n, both u_n and u_{n-1} vanish, we get

$$-\int_{A_n} (u_{n-1} - u_n)^2 \, dx = -\int_\Omega (u_{n-1} - u_n)^2 \, dx.$$

Returning to (5.8), we get the estimate for the time-discretized term:

$$[\text{right-hand side of (5.8)}] \leq \frac{1}{h^2} \int_\Omega [(u_{n-1} - u_{n-2})^2 - (u_{n-1} - u_n)^2] \, dx.$$

Combining this result and the gradient term estimate (5.7), we obtain

$$\int_\Omega \left[\frac{1}{h^2}(u_{n-1} - u_{n-2})^2 - \frac{1}{h^2}(u_n - u_{n-1})^2 + \frac{1}{2}|\nabla u_{n-2}|^2 - \frac{1}{2}|\nabla u_n|^2 \right] dx \geq 0.$$

Summing over $n = k, \ldots, k + \ell$, for any $1 \leq k < k + \ell \leq N$, we arrive at

$$\int_\Omega \left[\frac{1}{h^2}(u_{k-1} - u_{k-2})^2 - \frac{1}{h^2}(u_{k+\ell} - u_{k+\ell-1})^2 \right.$$
$$\left. + \frac{1}{2}|\nabla u_{k-2}|^2 + \frac{1}{2}|\nabla u_{k-1}|^2 - \frac{1}{2}|\nabla u_{k+\ell-1}|^2 - \frac{1}{2}|\nabla u_{k+\ell}|^2 \right] dx \geq 0,$$

which means $E_{k-1} \geq E_{k+\ell}$, and thanks to arbitrariness of k and ℓ implies (5.4).

□

5.3 Approximate Weak Solution

Minimizers of J_n^{CN} have the following subsolution property.

Proposition 5.1 *Any minimizer u of J_n^{CN} satisfies the following inequality for arbitrary nonnegative $\zeta \in H_0^1(\Omega)$:*

$$\int_{\Omega \cap \mathcal{S}_n(u)} \frac{u - 2u_{n-1} + u_{n-2}}{h^2} \zeta \, dx + \int_\Omega \nabla \frac{u + u_{n-2}}{2} \cdot \nabla \zeta \, dx \leq 0. \qquad (5.9)$$

Proof Fixing $\zeta \in C_0^\infty(\Omega)$ with $\zeta \geq 0$, we have for any $\varepsilon > 0$,

$$0 \leq J_n^{CN}(u - \varepsilon\zeta) - J_n^{CN}(u) \quad \text{(by the minimality of } u\text{)}$$

$$= \int_\Omega \frac{|(u - \varepsilon\zeta) - 2u_{n-1} + u_{n-2}|^2}{2h^2} \chi_{S_n(u-\varepsilon\zeta)} \, dx + \frac{1}{4} \int_\Omega |\nabla(u - \varepsilon\zeta) + \nabla u_{n-2}|^2 \, dx$$

$$- \left(\int_\Omega \frac{|u - 2u_{n-1} + u_{n-2}|^2}{2h^2} \chi_{S_n(u)} \, dx + \frac{1}{4} \int_\Omega |\nabla u + \nabla u_{n-2}|^2 \, dx \right).$$

$$(5.10)$$

Noting that

$$\chi_{S_n(u-\varepsilon\zeta)} - \chi_{S_n(u)} \leq 0,$$

$$|(u - \varepsilon\zeta) - 2u_{n-1} + u_{n-2}|^2 - |u - 2u_{n-1} + u_{n-2}|^2$$

$$= -2\varepsilon\zeta(u - 2u_{n-1} + u_{n-2}) + \varepsilon^2\zeta^2,$$

$$|\nabla(u - \varepsilon\zeta) + \nabla u_{n-2}|^2 - |\nabla u + \nabla u_{n-2}|^2 = -2\varepsilon(\nabla u + \nabla u_{n-2}) \cdot \nabla\zeta + \varepsilon^2|\nabla\zeta|^2,$$

we continue the estimate as

$$(5.10) \leq \int_\Omega \left(-\frac{\varepsilon}{h^2}\zeta(u - 2u_{n-1} + u_{n-2}) + \frac{\varepsilon^2}{2h^2}\zeta^2 \right) \chi_{S_n(u)} \, dx$$

$$+ \int_\Omega \left(-\frac{\varepsilon}{2}(\nabla u + \nabla u_{n-2}) \cdot \nabla\zeta + \frac{\varepsilon^2}{4}|\nabla\zeta|^2 \right) dx.$$

Dividing by ε, letting ε decrease to zero, and applying a density argument concludes the proof. □

The following theorem is obtained by a standard argument of elliptic regularity theory but we shall briefly demonstrate it for the sake of completeness.

Theorem 5.3 *Suppose that u_{-1}, u_0 are nonnegative and belong to $L^\infty(\Omega) \cap C_{\text{loc}}^{0,\alpha_0}(\Omega)$ for some $\alpha_0 \in (0, 1)$. Then for every compact subset $\widetilde{\Omega} \Subset \Omega$, there is a positive constant $\alpha \in (0, 1)$ independent of n, such that minimizers $\widetilde{u}_1, \widetilde{u}_2, \ldots, \widetilde{u}_N$ belong to $C^{0,\alpha}(\widetilde{\Omega})$.*

To prove this, we prepare two lemmas.

Lemma 5.1 *There is a constant M, which may depend on h, such that*

$$\|\widetilde{u}_n\|_{L^\infty(\Omega)} \leq M \quad \text{for every } n = 1, 2, \ldots, N.$$

Proof We use mathematical induction over $n \geq 1$. First, for $n = 1$, we set $\psi_\delta := \widetilde{u}_1 - \delta(\widetilde{u}_1 + u_{-1} - k)^+ \in \mathcal{K}$, where $(\widetilde{u}_1 + u_{-1} - k)^+ := \max\{\widetilde{u}_1 + u_{-1} - k, 0\}, \delta > 0$, and $k \geq 1$. We calculate the quantity $J_1^{CN}(\psi_\delta) - J_1^{CN}(\widetilde{u}_1)$, which is nonnegative by the minimality of $\widetilde{u}_1$. Noting that $\mathcal{S}_1(\psi_\delta) \subset \mathcal{S}_1(\widetilde{u}_1)$, we have

$$0 \leq J_1^{CN}(\psi_\delta) - J_1^{CN}(\widetilde{u}_1)$$

$$\leq \int_\Omega \left(\frac{|\psi_\delta - 2u_0 + u_{-1}|^2}{2h^2} - \frac{|\widetilde{u}_1 - 2u_0 + u_{-1}|^2}{2h^2} \right) \chi_{\mathcal{S}_1(\widetilde{u}_1)} \, dx$$

$$+ \frac{1}{4} \int_\Omega \left(|\nabla\psi_\delta + \nabla u_{-1}|^2 - |\nabla\widetilde{u}_1 + \nabla u_{-1}|^2 \right) dx.$$

Dividing by δ, letting $\delta \to 0+$, and setting $A_k := \{\widetilde{u}_1 + u_{-1} > k\}$, we get

$$0 \leq - \int_{A_k \cap \mathcal{S}_1(\widetilde{u}_1)} \frac{\widetilde{u}_1 - 2u_0 + u_{-1}}{h^2} (\widetilde{u}_1 + u_{-1} - k) \, dx - \frac{1}{2} \int_{A_k} |\nabla\widetilde{u}_1 + \nabla u_{-1}|^2 \, dx$$

$$\leq \int_{A_k \cap \mathcal{S}_1(\widetilde{u}_1)} \frac{2u_0}{h^2} (\widetilde{u}_1 + u_{-1} - k) \, dx - \frac{1}{2} \int_{A_k} |\nabla\widetilde{u}_1 + \nabla u_{-1}|^2 \, dx$$

$$\leq C \left(\frac{1}{2} \int_{A_k} (\widetilde{u}_1 + u_{-1} - k)^2 \, dx + \frac{1}{2}|A_k| \right) - \frac{1}{2} \int_{A_k} |\nabla\widetilde{u}_1 + \nabla u_{-1}|^2 \, dx,$$

where C is a constant depending on h and L^∞-norm of u_0. Since $k \geq 1$, we get

$$\int_{A_k} |\nabla\widetilde{u}_1 + \nabla u_{-1}|^2 \, dx \leq C \left(\int_{A_k} (\widetilde{u}_1 + u_{-1} - k)^2 \, dx + k^2 |A_k| \right).$$

Therefore, by Ladyzhenskaya and Uraltseva [38, Theorem 2.5.1], we find that $\widetilde{u}_1 + u_{-1} \in L^\infty(\Omega)$, and hence $\widetilde{u}_1 \in L^\infty(\Omega)$.

Next, we assume that $\widetilde{u}_k \in L^\infty(\Omega)$ for all $k = 1, \ldots, n-1$. Since $u_k = \max\{\widetilde{u}_k, 0\}$ belongs to $L^\infty(\Omega)$ for all $k = 1, \ldots, n-1$, by repeating the above argument with $\widetilde{u}_1, u_0, u_{-1}$ replaced by $\widetilde{u}_n, u_{n-1}, u_{n-2}$, respectively, we get $\widetilde{u}_n + u_{n-2} \in L^\infty(\Omega)$. Therefore, $\widetilde{u}_n \in L^\infty(\Omega)$. The bound depends only on Ω, u_0, v_0, h but not on n. $\square$

Lemma 5.2 *Fix $d > 0$. There exists $\gamma = \gamma(\Omega, M, d, h) > 0$ such that for $U = \pm(\widetilde{u}_n + u_{n-2})$,*

$$\int_{A_{k,r-\sigma r}} |\nabla U|^2 \, dx \leq \gamma \left[\frac{1}{(\sigma r)^2} \sup_{B_r} (U - k)^2 + 1 \right] |A_{k,r}|$$

for all $\sigma \in (0,1)$, $B_r \subset \Omega$, and k with $k \geq \max_{B_r} U - d$, where $A_{k,r} := \{x \in B_r; \ U(x) > k\}$, and B_r is a ball of radius r.

Proof For fixed $n \geq 1$, first we show the statement for $U = \widetilde{u}_n + u_{n-2}$. We set $\zeta = \eta^2 \max\{\widetilde{u}_n + u_{n-2} - k, 0\}$ in Proposition 5.1, where k is a real number with $k \geq \max_{B_r}(\widetilde{u}_n + u_{n-2}) - d$; further, η is a smooth function with $\operatorname{spt} \eta \subset B_r$, $0 \leq \eta \leq 1$, $\eta \equiv 1$ in B_s, $|\nabla\eta| \leq 2/(r-s)$ in $B_r \setminus B_s$, and $s = r - \sigma r \in (0, r)$,

$\sigma \in (0, 1)$. Then, using the boundedness of $\tilde{u}_n, u_{n-1}, u_{n-2}$, we get

$$0 \leq - \int_{A_{k,r} \cap S_n(\tilde{u}_n)} \frac{\tilde{u}_n - 2u_{n-1} + u_{n-2}}{h^2} \eta^2 (\tilde{u}_n + u_{n-2} - k)\, dx$$

$$- \int_{A_{k,r}} (\nabla \tilde{u}_n + \nabla u_{n-2}) \cdot (\eta \nabla \eta)(\tilde{u}_n + u_{n-2} - k)\, dx$$

$$- \frac{1}{2} \int_{A_{k,r}} |\nabla \tilde{u}_n + \nabla u_{n-2}|^2 \eta^2\, dx$$

$$\leq C |A_{k,r}| + \frac{1}{4} \int_{A_{k,r}} |\nabla(\tilde{u}_n + u_{n-2})|^2 \eta^2\, dx + \int_{A_{k,r}} |\nabla \eta|^2 (\tilde{u}_n + u_{n-2} - k)^2\, dx$$

$$- \frac{1}{2} \int_{A_{k,r}} |\nabla(\tilde{u}_n + u_{n-2})|^2 \eta^2\, dx$$

$$\leq C \left(1 + \frac{1}{(\sigma r)^2} \sup_{B_r}(\tilde{u}_n + u_{n-2} - k)^2\right) |A_{k,r}| - \frac{1}{4} \int_{A_{k,s}} |\nabla(\tilde{u}_n + u_{n-2})|^2\, dx,$$

where the constant C depends only on h, M, d, Ω.

Next, we prove the same inequality for $U = -(\tilde{u}_n + u_{n-2})$. Note that $-\tilde{u}_n$ is a minimizer of the functional

$$J_n^{CN-}(w) := \int_{\Omega \cap S_n^-(w)} \frac{|w + 2u_{n-1} - u_{n-2}|^2}{2h^2}\, dx + \frac{1}{4} \int_{\Omega} |\nabla w - \nabla u_{n-2}|^2\, dx$$

in $\mathcal{K}^- = H_0^1(\Omega)$, where $S_n^-(w) = \{w < 0\} \cup \{u_{n-1} > 0\} \cup \{u_{n-2} > 0\}$.

Now, we set $\varphi := -\tilde{u}_n - \zeta \in \mathcal{K}^-$ where $\zeta := \eta \max\{-\tilde{u}_n - u_{n-2} - k, 0\}$, k is a real number with $k \geq \max_{B_r}(-\tilde{u}_n - u_{n-2}) - d$, and η is a smooth function chosen in the same way as above. Then, by the minimality of $-\tilde{u}_n$,

$$0 \leq J_n^{CN-}(\varphi) - J_n^{CN-}(-\tilde{u}_n)$$

$$\leq \int_{\Omega \cap S_n^-(\varphi)} \frac{2(-\tilde{u}_n + 2u_{n-1} - u_{n-2})\zeta + \zeta^2}{2h^2}\, dx$$

$$+ \int_{\Omega} \frac{|\tilde{u}_n - 2u_{n-1} + u_{n-2}|^2}{2h^2} \left(\chi_{S_n^-(\varphi)} - \chi_{S_n^-(-\tilde{u}_n)}\right) dx$$

$$+ \frac{1}{4} \int_{\Omega} |\nabla \varphi - \nabla u_{n-2}|^2\, dx - \frac{1}{4} \int_{\Omega} |-\nabla \tilde{u}_n - \nabla u_{n-2}|^2\, dx. \qquad (5.11)$$

Note that the term in the third line is less than or equal to $1/(2h^2) \int_{\mathrm{spt}\, \zeta} |\tilde{u}_n - 2u_{n-1} + u_{n-2}|^2\, dx$, since $\chi_{S_n^-(\varphi)} - \chi_{S_n^-(-\tilde{u}_n)}$ can be positive only for x satisfying $0 \leq -\tilde{u}_n(x) < \zeta(x)$. Therefore, recalling that $\mathrm{spt}\, \zeta \subset A_{k,r}$, the first two terms on the right-hand side of (5.11) are less than or equal to $C|A_{k,r}|$, where C is a constant depending only Ω, M, d, h. Then, we continue the estimate of (5.11) as follows:

$$0 \leq C|A_{k,r}| + \frac{1}{2}\int_{A_{k,r}}(-\tilde{u}_n - u_{n-2} - k)^2|\nabla\eta|^2\,dx$$

$$+ \frac{1}{2}\int_{A_{k,r}}(1-\eta)^2|-\nabla\tilde{u}_n - \nabla u_{n-2}|^2\,dx - \frac{1}{4}\int_{A_{k,r}}|-\nabla\tilde{u}_n - \nabla u_{n-2}|^2\,dx$$

$$\leq C|A_{k,r}| + \frac{2}{(\sigma r)^2}\int_{A_{k,r}}(-\tilde{u}_n - u_{n-2} - k)^2\,dx$$

$$+ \frac{1}{2}\int_{A_{k,r}}|\nabla(-\tilde{u}_n - u_{n-2})|^2\,dx - \frac{3}{4}\int_{A_{k,s}}|\nabla(-\tilde{u}_n - u_{n-2})|^2\,dx.$$

Therefore, we get

$$\int_{A_{k,s}}|\nabla U|^2\,dx \leq C|A_{k,r}| + \theta\int_{A_{k,r}}|\nabla U|^2\,dx + \frac{8}{3}\frac{1}{(\sigma r)^2}\int_{A_{k,r}}(U-k)^2\,dx,$$

where $\theta = \frac{2}{3} < 1$. By Lemma V. 3.1 in [24], we obtain

$$\int_{A_{k,s}}|\nabla U|^2\,dx \leq C|A_{k,r}| + \frac{8}{3}\frac{1}{(\sigma r)^2}\int_{A_{k,r}}(U-k)^2\,dx,$$

which is the desired estimate for $U = -(\tilde{u}_n + u_{n-2})$. □

Proof (of Theorem 5.3) Lemmas 5.1 and 5.2 imply that $\tilde{u} := \tilde{u}_n + u_{n-2}$ for $n \geq 1$ belong to the De Giorgi class $\mathcal{B}_2(\Omega, M, \gamma, d)$ (see Definition 4.3). Thus, by De Giorgi's embedding theorem [38, Section 2.6], $\tilde{u}_n + u_{n-2} \in C^{0,\tilde{\alpha}}(\tilde{\Omega})$ for some $\tilde{\alpha} \in (0, 1)$ which is independent of n. We can now prove that $\tilde{u}_n \in C^{0,\alpha}(\tilde{\Omega})$ for some $\alpha \in (0, 1)$. To this end, set $\alpha := \min\{\alpha_0, \tilde{\alpha}\}$. For $n = 1$, by the fact that $\tilde{u}_1 + u_{-1} \in C^{0,\tilde{\alpha}}(\tilde{\Omega}) \subset C^{0,\alpha}(\tilde{\Omega})$, and the assumption $u_{-1} \in C^{0,\alpha_0}(\tilde{\Omega}) \subset C^{0,\alpha}(\tilde{\Omega})$, we see that $\tilde{u}_1 \in C^{0,\alpha}(\tilde{\Omega})$. Hence, $u_1 = \max\{\tilde{u}_1, 0\}$ belongs to the same space. Now, we assume that $\tilde{u}_k \in C^{0,\alpha}(\tilde{\Omega})$ for all $k = 1, \ldots, n-1$. Then, since $u_{n-2} \in C^{0,\alpha}(\tilde{\Omega})$, and $\tilde{u}_n + u_{n-2} \in C^{0,\tilde{\alpha}}(\tilde{\Omega}) \subset C^{0,\alpha}(\tilde{\Omega})$, we have $\tilde{u}_n \in C^{0,\alpha}(\tilde{\Omega})$. □

By the above theorem, we can choose the support of test functions within the open set $\{\tilde{u}_n > 0\}$, which leads to the following first variation formula for J_n^{CN}.

Proposition 5.2 *Any minimizer u of J_n^{CN} for $n = 1, 2, \ldots, N$, satisfies the identity*

$$\int_{\Omega}\left(\frac{u - 2u_{n-1} + u_{n-2}}{h^2}\phi + \nabla\frac{u + u_{n-2}}{2} \cdot \nabla\phi\right)dx = 0 \qquad (5.12)$$

for all $\phi \in C_0^{\infty}(\Omega \cap \{u > 0\})$.

Proof Since $\{u > 0\}$ is an open set by Theorem 5.3, we can calculate the first variation of J_n^{CN} using $u + \varepsilon\phi$ with $\phi \in C_0^{\infty}(\Omega \cap \{u > 0\})$ as a test function. The result then follows by noting that there exists $\varepsilon_0 > 0$ such that $\chi_{S_n(u+\varepsilon\phi)} = \chi_{S_n(u)}$ for $|\varepsilon| < \varepsilon_0$. $\square$

The strong form of Eq. (5.12) is

$$\frac{u - 2u_{n-1} + u_{n-2}}{h^2} = \Delta\left(\frac{u + u_{n-2}}{2}\right),$$

and thus we sometimes say that the functional (5.1) is of *Crank–Nicolson type*.

Our next goal is to construct a weak solution. Accordingly, we carry out interpolation in time of the cut-off minimizers $\{u_n\}$ of J_n^{CN}, and introduce approximate weak solutions. In particular, we define $\bar{u}^h$ and u^h by

$$\bar{u}^h(x, t) = u_n(x), \qquad n = -1, 0, \ldots, N$$

$$u^h(x, t) = \frac{t - (n - 1)h}{h} u_n(x) + \frac{nh - t}{h} u_{n-1}(x), \qquad n = 0, 1, \ldots, N,$$

for $(x, t) \in \Omega \times ((n-1)h, nh]$. These functions allow us to construct an approximate solution satisfying

$$\int_0^T \int_\Omega \left(\frac{u_t^h(t) - u_t^h(t - h)}{h}\phi + \nabla\frac{\bar{u}^h(t) + \bar{u}^h(t - 2h)}{2} \cdot \nabla\phi\right) dx\, dt = 0$$

$$(5.13)$$

for all $\phi \in C_0^\infty(\Omega \times [0, T) \cap \{u^h > 0\})$, and

$$u^h \equiv 0 \quad \text{in} \quad \Omega \times (0, T) \setminus \{u^h > 0\}.$$

If one can pass to the limit as $h \to 0+$, then the above approximate weak solutions are expected to converge to a weak solution of Definition 4.1.

Lemma 5.3 *There exists a decreasing sequence $\{h_j\}_{j=1}^\infty$ with $h_j \to 0+$ (denoted as h again) and $u \in H^1(0, T; L^2(\Omega)) \cap L^\infty(0, T; H_0^1(\Omega))$ such that*

$$u_t^h \rightharpoonup u_t \qquad \text{weakly} * \text{in } L^\infty(0, T; L^2(\Omega)), \qquad (5.14)$$

$$\nabla\bar{u}^h \rightharpoonup \nabla u \qquad \text{weakly} * \text{in } L^\infty(0, T; L^2(\Omega)), \qquad (5.15)$$

$$u^h \to u \qquad \text{strongly in } L^2(Q_T). \qquad (5.16)$$

When in addition $\dim(\Omega) = 1$ *then*

$$u_h \rightrightarrows u \quad \text{uniformly on } [0, T) \times \Omega. \qquad (5.17)$$

Proof Since the estimate (5.4) implies

$$\|u_t^h(t)\|_{L^2(\Omega)}^2 + \|\nabla \overline{u}^h(t)\|_{L^2(\Omega)}^2 \leq C \quad \text{for a.e. } t \in (0, T), \tag{5.18}$$

the existence of a convergent subsequence is proved exactly in the same way as in Lemma 4.1. □

We are now in the same situation as in Sect. 4.2: we have constructed a sequence of approximate solutions converging to a limit. In order to prove that this limit is a weak solution of the original problem in the sense of Definition 4.1, uniform convergence of approximate solutions is sufficient, but this kind of convergence is obtained only for the one-dimensional case. We will conclude by taking the limit in this special case of spatial dimension one. To this end, the following lemma is needed.

Lemma 5.4 *Define* $\overline{w}^h(x, t) := \overline{u}^h(t - 2h)$ *for* $t \in (0, T)$. *Then,*

$$\nabla \overline{w}^h \rightharpoonup \nabla u \quad \text{weakly} * \text{ in } L^\infty(0, T; L^2(\Omega)).$$

Proof We omit the space variable x for simplicity. Let us fix $U \in L^1(0, T; L^2(\Omega))$ and extend it by zero outside of $(0, T)$. The extended function, denoted again by U, belongs to $L^1(-\infty, \infty; L^2(\Omega))$. We calculate as follows:

$$\left| \int_0^T \langle \nabla \overline{w}^h(t), U(t) \rangle_{L^2(\Omega)} \, dt - \int_0^T \langle \nabla u(t), U(t) \rangle_{L^2(\Omega)} \, dt \right|$$

$$= \left| \int_{-2h}^{T-2h} \langle \nabla \overline{u}^h(t), U(t + 2h) \rangle \, dt - \int_0^T \langle \nabla u(t), U(t) \rangle \, dt \right|$$

$$\leq \left| \int_0^{T-2h} \langle \nabla \overline{u}^h(t), U(t + 2h) - U(t) \rangle \, dt \right| + \left| \int_0^T \langle \nabla \overline{u}^h(t) - \nabla u(t), U(t) \rangle \, dt \right|$$

$$+ \left| \int_{T-2h}^T \langle \nabla \overline{u}^h(t), U(t) \rangle \, dt \right| + \left| \int_{-2h}^0 \langle \nabla \overline{u}^h(t), U(t + 2h) \rangle \, dt \right|$$

$$\leq C \int_{-\infty}^\infty \|U(t + 2h) - U(t)\|_{L^2(\Omega)} \, dt + \left| \int_0^T \langle \nabla \overline{u}^h(t) - \nabla u(t), U(t) \rangle \, dt \right|$$

$$+ C \int_{T-2h}^T \|U(t)\|_{L^2(\Omega)} \, dt + C \int_0^{2h} \|U(t)\|_{L^2(\Omega)} \, dt, \tag{5.19}$$

where the constant C is independent of h. Letting $h \to 0+$, the second term converges to 0 by (5.15), and the remaining terms vanish thanks to the integrability of U. □

We can now state the existence result.

Theorem 5.4 *Let Ω be a bounded domain in $\mathbb{R}$. Then the problem to find a stationary point of the functional (4.2) with $F \equiv 0$ in the set $\mathcal{K}^T$ (given by (4.4)) has a weak solution in the sense of Definition 4.1.*

Proof The task is to take $h \to 0+$ in (5.13). For the time-discrete term we can proceed exactly in the same way as in Sect. 4.3, while for the gradient term we have in view of Lemma 5.4 and (5.15),

$$\int_0^T \int_\Omega \nabla \frac{\overline{u}^h(t) + \overline{u}^h(t - 2h)}{2} \cdot \nabla \phi \, dx \, dt = \int_0^T \int_\Omega \nabla \frac{\overline{u}^h(t) + \overline{w}^h(t)}{2} \cdot \nabla \phi \, dx \, dt$$

$$\to \int_0^T \int_\Omega \nabla u \cdot \nabla \phi \, dx \, dt.$$

Moreover, $u \geq 0$ and $u(t) \in H_0^1(\Omega)$ for a.e. t. This implies that u is a weak solution and the proof is finished.   □

Note that the fact that (4.31) holds for any test function ϕ compactly supported in the support of the limit function u is a consequence of the uniform convergence of the approximating sequence $\{u^h\}$, which was in turn obtained through an embedding and the Arzelà–Ascoli theorem. However, this embedding is true only in spatial dimension 1, preventing us from extending the existence result to higher dimensions. The papers [11, 12] succeed in proving existence in arbitrary dimension by introducing a different (weaker) definition of a weak solution. Nevertheless, an important aspect of our strategy, common to that in [11, 12], is that a uniform estimate is available for the approximate solutions and hence a unique limit function can be identified. How much can be said about this limit remains an open question. On the other hand, the energy-preserving property of our approximation method can be exploited not only in building robust numerical algorithms, but also in proving uniqueness of solutions for certain wave-type problems. The existing variational approach was not successful in proving uniqueness, and hence investigating the new scheme from this viewpoint presents an interesting direction for future research.

5.4   Numerical Results

In order to visually illustrate the performance of the energy-preserving scheme, we present comparison of the results obtained by the original and the improved discrete Morse flow for the problem of finding stationary points of (4.2). Namely, we minimize the Crank–Nicolson-type functional

$$J_n(u) = \int_{\Omega \cap (\{u>0\} \cup \{u_{n-1}>0\} \cup \{u_{n-2}>0\})} \frac{|u - 2u_{n-1} + u_{n-2}|^2}{2h^2} \, dx$$

$$+ \frac{1}{4} \int_\Omega |\nabla u + \nabla u_{n-2}|^2 \, dx,$$

and compare the output with results due to the original discrete Morse flow method presented in [66], which uses the functional

$$\mathcal{J}_n(u) = \int_{\Omega \cap (\{u > 0\} \cup \{u_{n-1} > 0\})} \frac{|u - 2u_{n-1} + u_{n-2}|^2}{2h^2}\, dx + \frac{1}{2} \int_\Omega |\nabla u|^2\, dx.$$

In the numerical calculation, we simply use the corresponding functionals without any restriction on the integration domain, subsequently truncate their minimizers $\widetilde{u}_n$ by setting $u_n := \max\{\widetilde{u}_n, 0\}$ and regard u_n as a numerical solution at time level $t = nh$. The minimization problems are discretized by piecewise linear finite elements and solved using a gradient descent method.

In the one-dimensional setting, this problem can describe the dynamics of a string hitting a plane with zero reflection constant. We would like to bring attention to several other approaches to solve similar problems in various settings: see, for example, [72, 73] for the perfectly elastic impact, [23] for another method of discretization, or to [87] for a computation having in mind applications to musical instruments.

We consider a string in the interval $\Omega = (0, 1)$, with the initial condition

$$u_0(x) = \begin{cases} \dfrac{5}{2}x + \dfrac{1}{5} & \text{if } 0 \le x < \dfrac{2}{5}, \\[3mm] -\dfrac{5}{3}(x - 1) + \dfrac{1}{5} & \text{otherwise}, \end{cases}$$

and $v_0 \equiv 0$. Figure 5.1 compares the numerical behavior of both schemes. The Crank–Nicolson method preserves sharp corners in the graph of the solution as time progresses, while this is not the case for the discrete Morse flow method, where corners are smoothed. The lower panel shows the behavior of the free boundary. In fact, when the string first hits the obstacle and detaches, energy is dissipated through the contact, and the string never touches the obstacle again during further oscillations. Therefore, we plot the boundary of the set $\{(t, x); u(x, t) < \varepsilon\}$ for a small $\varepsilon > 0$.

The difference in values of the energy (3.30) is made clear in Fig. 5.2. Energy dissipates when the string touches the obstacle, while it should be preserved before and after the contact of the string with the obstacle. This is realized by the Crank–Nicolson scheme but in discrete Morse flow the energy decays even during the non-contact stage. The difference between both methods becomes even more prominent when numerically solving problems in higher spatial dimensions [3].

To further test the energy decay tendency of both methods, we solved the problem without free boundary with the initial condition $u_0 = \sin(2n\pi x)$ and $v_0 \equiv 0$. As shown in Fig. 5.3 (taken from [3]), energy of the original discrete Morse flow decays more rapidly with decreasing time resolution and increasing wave frequency, while the Crank–Nicolson scheme preserves energy independently of these parameters.

In two dimensions, the volume-constrained problem can model the motion of droplets over a flat surface (see, e.g., [26]). We simulated such a problem by

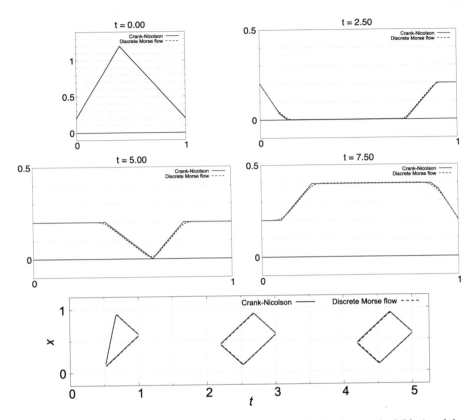

Fig. 5.1 Numerical solution at four distinct times for the Crank–Nicolson method (blue) and the original discrete Morse flow method (red). Lower panel shows corresponding evolution of the free boundary. Discretization parameters are $h = \Delta x = 1.0 \times 10^{-4}$

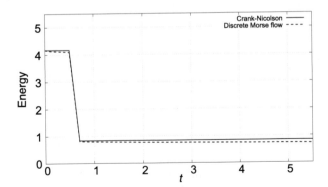

Fig. 5.2 Evolution of energy of the numerical solutions from Fig. 5.1

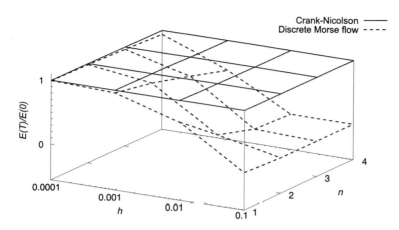

Fig. 5.3 Comparison of energy decay tendency for a problem without free boundary under initial data $u_0 = \sin(2n\pi x)$ and $v_0 \equiv 0$. Spatial grid size is taken equal to time step h

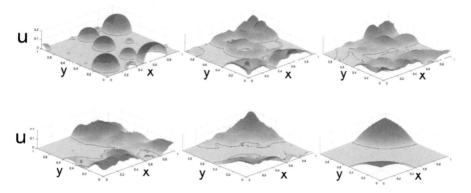

Fig. 5.4 Simulation of droplet motion (time increases from top to bottom, left to right)

solving a minimization problem where volume and nonnegativity constraints are added to the functionals by means of indicator functions. An example of numerical simulation is shown in Fig. 5.4. Starting from the initial condition prescribed as a collection of spherical caps, the droplets oscillate while coalescing into larger groups.

Chapter 6
Numerical Examples and Applications

In this chapter, we will discuss computational aspects of the discrete Morse flow (DMF), and illustrate three applications. We begin by showing how functional values can be approximated by means of the finite element method. Methods for performing the functional minimizations are also briefly discussed.

In our first application, we show how the DMF can be combined with a threshold dynamics algorithm, known as the HMBO, to simulate interfacial motion by volume-preserving hyperbolic mean curvature flow [27]. The PDE used by the threshold dynamics algorithm is formally shown to be a hyperbolic free boundary problem, and we investigate the oscillatory motion of the interface. In our second application, the DMF is used to model and simulate an elastodynamic contact problem [2], and in our final application, we show how the DMF can be used in simulating the impact of an elastic shell with an obstacle [32].

6.1 Finite Element Approximation

Using the discrete Morse flow in computations requires one to approximate, and then compute functional values. In turn, standard optimization algorithms enable one to obtain computational realizations of functional minimizers. In this light, we will explain the numerical implementation of the discrete Morse flow. In addition, we will also show how the variational nature of this method also allows one to easily incorporate constraints in the minimization process, via penalty terms.

First, we explain computational aspects of the DMF for functionals with the form:

$$\mathcal{F}_n(u) = \int_\Omega \left(\frac{|u - 2u_{n-1} + u_{n-2}|^2}{2h^2} + \frac{1}{2}|\nabla u|^2 \right) dx + \frac{1}{\varepsilon}\mathcal{P}(u),$$

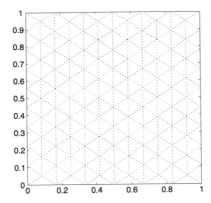
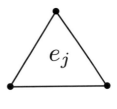

Fig. 6.1 (Left) A rough triangulation of the unit square. (Right) A typical element e_j with three vertices

where $h > 0$ is the length of the time step, $\boldsymbol{u}_{n-1}, \boldsymbol{u}_{n-2}$ are given vector-valued functions

$$\boldsymbol{u}_{n-1}, \boldsymbol{u}_{n-2} : \Omega \to \mathbb{R}^K,$$

and $\varepsilon > 0$ is a parameter for controlling the influence of a given penalty term $\mathcal{P}(\boldsymbol{u})$. For the sake of clarity, we will assume Ω is a given domain in $\mathbb{R}^2$, but we remark that this approach carries over to other functionals and general dimensions.

The finite element method approach to approximating functional values begins by discretizing Ω into a finite set of triangular elements $\{e_j\}_{j=1}^M$. This can be done in a number of ways, for example by using a Delaunay triangulation. Figure 6.1 shows a rough triangulation of the unit square, together with a typical element. The values of the unknown function $\boldsymbol{u} : \Omega \to \mathbb{R}^K$ are prescribed at certain nodes in the triangulation and an assumption about their extension to each element is made. We will use the so-called $\mathbf{P}^1$-approximation, that is, over each element e_j, the function $\boldsymbol{u}$ is assumed to be a linear interpolation of the three values at each of the vertices of the element:

$$\boldsymbol{u}|_{e_j} \approx \boldsymbol{u}^j(x, y) := \boldsymbol{\alpha}^j x + \boldsymbol{\beta}^j y + \boldsymbol{\gamma}^j, \qquad (x, y) \in e_j, \tag{6.1}$$

where components of the vectors

$$\boldsymbol{\alpha}^j = (\alpha_1^j, \alpha_2^j, \cdots, \alpha_K^j)^T, \quad \boldsymbol{\beta}^j = (\beta_1^j, \beta_2^j, \cdots, \beta_K^j)^T, \quad \boldsymbol{\gamma}^j = (\gamma_1^j, \gamma_2^j, \cdots, \gamma_K^j)^T$$

are determined as the solutions of the systems of equations

$$
\begin{pmatrix} x_1^j & y_1^j & 1 \\ x_2^j & y_2^j & 1 \\ x_3^j & y_3^j & 1 \end{pmatrix} \begin{pmatrix} \alpha_k^j \\ \beta_k^j \\ \gamma_k^j \end{pmatrix} = \begin{pmatrix} u_k(x_1^j, y_1^j) \\ u_k(x_2^j, y_2^j) \\ u_k(x_3^j, y_3^j) \end{pmatrix}, \qquad k = 1, 2, \ldots, K.
$$

Here the coordinates of the element vertices are expressed as (x_i^j, y_i^j), $i = 1, 2, 3$.

The value of the functional can then be approximated by accumulating values over each of the elements:

$$
\mathcal{F}_n(\boldsymbol{u}) \approx \sum_{j=1}^{M} \left(\int_{e_j} \left(\frac{\left| \boldsymbol{u}^j - 2\boldsymbol{u}_{n-1}^j + \boldsymbol{u}_{n-2}^j \right|^2}{2h^2} + \frac{1}{2} \left| \nabla \boldsymbol{u}^j \right|^2 \right) dx + \frac{1}{\varepsilon} \mathcal{P}(\boldsymbol{u}^j) \right),
$$

where M is the number of elements in the triangulation, and $\boldsymbol{u}^j$ denotes the $\mathbf{P}^1$-approximation (6.1) of $\boldsymbol{u}$ over element e_j. Straightforward calculations yield the following formula for approximating functional values:

$$
\int_{e_j} \frac{\left| \boldsymbol{u}^j - 2\boldsymbol{u}_{n-1}^j + \boldsymbol{u}_{n-2}^j \right|^2}{2h^2} \, dx = \int_{e_j} \sum_{k=1}^{K} \frac{\left| u_k^j - 2u_{k,n-1}^j + u_{k,n-2}^j \right|^2}{2h^2} \, dx
$$

$$
= \frac{|e_j|}{12h^2} \sum_{k=1}^{K} \sum_{i=1}^{3} \sum_{i'=1}^{3} w_k(x_i^j, y_i^j) w_k(x_{i'}^j, y_{i'}^j),
$$

$$
(6.2)
$$

where $w_k(x, y) := u_k(x, y) - 2u_{k,n-1}(x, y) + u_{k,n-2}(x, y)$ and $|e_j|$ denotes the area of the element e_j. Similarly,

$$
\int_{e_j} \frac{1}{2} \left| \nabla \boldsymbol{u}^j \right|^2 dx = \frac{1}{2} \int_{e_j} \sum_{k=1}^{K} |\nabla u_k^j|^2 \, dx = \frac{|e_j|}{2} \sum_{k=1}^{K} \left((\alpha_k^j)^2 + (\beta_k^j)^2 \right). \qquad (6.3)
$$

The approximation of the penalty term naturally depends on its specific formulation. For example, the droplet model computations in Fig. 5.4 incorporate the contact energy of a scalar function u:

$$
\mathcal{P}(u) = \int_{\Omega} C_s(x, y) \chi_{\{u>0\}}(x, y) \, dx \, dy,
$$

where $C_s(x, y)$ is a nonnegative weight function. Similarly, we note that penalty terms can be used to incorporate constraints. For example, in the case of volume

Fig. 6.2 Location of a free
boundary across an element

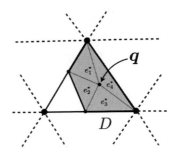

preservation, one could introduce a penalty term as follows:

$$\mathcal{P}(u) = \left(V_0 - \int_\Omega u \chi_{\{u>0\}}(x, y)\, dx\, dy \right)^2,$$

where the value of V_0 prescribes a target volume.

When the free boundary $\partial\{u > 0\}$ crosses through an element in the $\mathbf{P}^1$ setting, the resulting region of $\{u > 0\}$ restricted to the element is a convex set. Let D denote this region, and q be its center of mass. Then, using q, D can be partitioned into a finite set of triangular elements as $D = \bigcup e_j^*$ (see Fig. 6.2). This allows one to approximate the values of the contact energy term by accumulating values over each element as follows:

$$\int_{e_j} C_s(x, y) \chi_{\{u>0\}}(x, y)\, dx\, dy$$

$$\approx \begin{cases} \frac{|e_j|}{3}(C_s(x_1^j, y_1^j) + C_s(x_2^j, y_2^j) + C_s(x_3^j, y_3^j)), & e_j \subset \{u > 0\} \\ 0, & e_j \cap \{u > 0\} = \emptyset \\ \sum_{e_j^* \subset D} \frac{|e_j^*|}{3}(C_s(x_1^{j*}, y_1^{j*}) + C_s(x_2^{j*}, y_2^{j*}) + C_s(x_3^{j*}, y_3^{j*})), & \text{otherwise,} \end{cases}$$

where the coordinates of vertices of each e_j^* are denoted by (x_i^{j*}, y_i^{j*}), $i = 1, 2, 3$.

Functional minimizations can then be achieved by any number of optimization techniques, including steepest descent, conjugate gradient, Newton's method, or the Nelder–Mead method. We refer the reader to the monograph [63] or any other book on numerical methods for optimization problems.

6.2 Examples of Applications

We briefly present three examples of applications of the discrete Morse flow method to numerical simulations of phenomena formulated as hyperbolic free boundary problems, directing the interested reader to the original papers [2, 27, 32] for more details.

6.2.1 Volume-Preserving Hyperbolic Mean Curvature Flow

The hyperbolic mean curvature flow (HMCF) is an oscillating interfacial motion [29]:

$$\frac{d^2 x}{dt^2} = -\kappa(x, t)n(x, t), \tag{6.4}$$

where κ denotes the mean curvature of the interface, and n its unit outer normal. We refer to [8, 9, 21, 22, 30] for a general theory of such a class of motions, connected to relativistic strings, where they are derived in the form of minimal submanifolds in Minkowski space as a singular limit of nonlinear wave equations.

Evolution by (6.4) can, for example, be approximated by means of the front tracking method [91]. Another approach is to use the threshold dynamics HMBO algorithm [27], which we now describe.

Let Γ denote an interface, described as the boundary of an open set $\Omega^+ \subset \mathbb{R}^2$:

$$\Gamma = \partial \Omega^+.$$

The HMBO encodes the interface as the zero level set of a signed distance function:

$$\Gamma = \{x \in \mathbb{R}^2 : d_\Gamma^\pm(x) = 0\}, \tag{6.5}$$

where

$$d_\Gamma^\pm(x) = \begin{cases} \inf_{y \in \Gamma} \|x - y\| & x \in \Omega^+ \\ -\inf_{y \in \Gamma} \|x - y\| & \text{otherwise.} \end{cases} \tag{6.6}$$

By setting a time step $\tau > 0$, and observing the motion of the zero level set solutions to the wave equations:

$$\begin{cases} u_{tt} = c^2 \Delta u & x \in \mathbb{R}^2,\ 0 < t < \tau, \\ u(x, t = 0) = u_0(x), & u_t(x, t = 0) = v_0(x), \end{cases} \tag{6.7}$$

one has the following result [27]:

Theorem 6.1 *Given two interfaces Γ^0 and Γ^{-1} (defined using the initial normal velocity field w_0 along Γ^0), for $k = 0, 1, 2, \ldots$ repeatedly evolve $u_0(x) := 2d_{\Gamma^k}^\pm - d_{\Gamma^{k-1}}^\pm$ by the wave equation (6.7) where $c^2 = 2$, with initial velocity $v_0 = 0$, defining Γ^{k+1} as the zero level set of the obtained solution at time τ. Then, up to an error of order $O(\tau)$, the family $\{\Gamma^k\}_{k=0,1,\ldots}$ realizes the evolution of Γ^0 with normal acceleration equal to $-\kappa$ and initial velocity w_0.*

By adding a volume constraint to the HMCF, one can investigate the geometric flow of soap film dynamics [29]. One constructs the signed distance function $d_{\Gamma_0}^{\pm}(x)$ and uses the initial velocity field along the interface to define $d_{\Gamma_{-1}}^{\pm}(x)$. Then the volume-preserving HMCF can be formally realized by adding a corresponding penalty to the hyperbolic DMF:

$$\mathcal{F}_n(u) = \int_{\Omega} \left(\frac{|u - 2u_{n-1} + u_{n-2}|^2}{2h^2} + \frac{1}{2}|\nabla u|^2 \right) dx + \frac{1}{\varepsilon}\mathcal{P}(u), \qquad (6.8)$$

where h is a substep of τ and, for a given parameter $\varepsilon > 0$, the penalty term can be given the form

$$\mathcal{P}(u) = \left(V_0 - \int_{\Omega} \chi_{\{u>0\}}\, dx \right)^2. \qquad (6.9)$$

Here V_0 is a positive number corresponding to the area enclosed by the interface at the initial time:

$$V_0 = \int_{\Omega} \chi_{\{u_0>0\}}\, dx. \qquad (6.10)$$

The Euler–Lagrange equation for (6.8) is

$$\frac{u - 2u_{n-1} + u_{n-2}}{h^2} = \Delta u + \lambda_n(u)\delta_{\partial\{u>0\}}, \qquad (6.11)$$

where δ_S denotes the delta distribution supported on a curve S, and the value of the Lagrange multiplier can be expressed as

$$\lambda_n(u) = \frac{2}{\varepsilon}\left(V_0 - \int_{\{u>0\}} dx \right). \qquad (6.12)$$

Therefore the corresponding threshold dynamics algorithm for volume-preserving HMCF is to solve the following hyperbolic free boundary problems for each time step k:

$$\begin{cases} u_{tt}^k = c^2 \Delta u^k & x \in \mathbb{R}^2 \backslash \partial\{u^k > 0\},\ 0 < t < \tau \\ [u^k] = 0 & 0 < t < \tau \\ [u_n^k] = -\lambda_k(u^k) & 0 < t < \tau \\ u^k(x, t = 0) = 2d_{\Gamma^k}^{\pm}(x) - d_{\Gamma^{k-1}}^{\pm}(x) \\ u_t^k(x, t = 0) = 0. \end{cases} \qquad (6.13)$$

In the above, $[\cdot]$ designates the jump in its quantity across the normal of the interface $\partial\{u^k > 0\}$. Thus $[u^k] = 0$ imparts continuity, while $[u_n^k] = -\lambda_k(u^k)$ specifies a

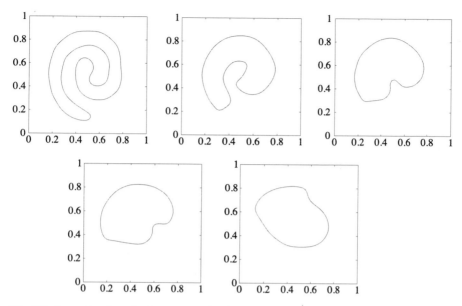

Fig. 6.3 Approximation of volume-preserving hyperbolic mean curvature flow (time is from left to right)

jump of magnitude $-\lambda_k$ in the normal derivative. Using the constrained DMF with the computational approach described above, we are able to examine the motion of the interfaces Γ^k which represent an approximation of volume-preserving HMCF.

Figure 6.3 shows a computation result where the initial condition is taken as the displayed spiral and initial velocity is set to zero. We observe the interface oscillate while conserving the area of its enclosure.

The algorithm can be extended, at least formally, to address multiphase problems by employing a vector-valued generalization of the signed distance function in the spirit of [27, 54]. Figure 6.4 shows a simulation of hyperbolic mean curvature flow

Fig. 6.4 Numerical solution of hyperbolic six-phase evolution associated to the length functional of interfacial network. The initial condition is shown by a thick line

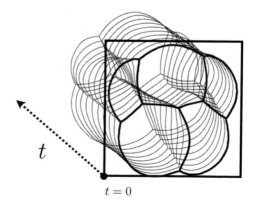

$t = 0$

for a network of interfaces separating six phases. Mathematical analysis of this type of multiphase interfacial evolution is an area for future development.

6.2.2   Elastodynamic Contact Problem

The DMF can also be used to treat hyperbolic free boundary problems arising in elastodynamics. Here we will show results of the DMF in modeling a rolling contact problem describing the motion of a rotating elastic body which comes into contact with a solid obstacle.

Let $\Omega \subset \mathbb{R}^2$ be a bounded domain representing a reference elastic body with boundary components Γ_C, Γ_D (see Fig. 6.5) which is located above an obstacle g. Here g is assumed to be a time-dependent nonnegative smooth function, $g : [0, T) \to \mathbb{R}$.

The model equation that we consider is then described by the following obstacle problem [2], evolving from an initial displacement ξ^0 and the initial velocity field η^0:

$$
\begin{cases}
\rho\ddot{\xi} - \operatorname{div}\sigma[\xi] = \rho f(\cdot, \cdot, \xi, \dot{\xi}) & \text{in } \Omega \times (0, T) \\
\xi = 0 & \text{on } \Gamma_D \times [0, T) \\
(I + \xi) \cdot (R^T(\theta)e_2) \geq g & \text{on } \Gamma_C \times [0, T) \\
(\sigma[\xi]n) \cdot (R^T(\theta)e_1) = 0 & \text{on } \Gamma_C \times [0, T) \\
(\sigma[\xi]n) \cdot (R^T(\theta)e_2) \geq 0 & \text{on } \Gamma_C \times [0, T) \\
((I + \xi) \cdot (R^T(\theta)e_2) - g)(\sigma[\xi]n) \cdot (R^T(\theta)e_2) = 0 & \text{on } \Gamma_C \times [0, T) \\
\xi(\cdot, 0) = \xi^0 & \text{in } \Omega \\
\dot{\xi}(\cdot, 0) = \eta^0 & \text{in } \Omega
\end{cases}
$$

$$(6.14)$$

where $\rho > 0, \theta : [0, T) \to \mathbb{R}$ is smooth and prescribes the rotation force acting on the elastic body, and f is an outer force defined by

$$
f(x, t, \xi, \dot{\xi}) = \ddot{\theta}(t)R(-\pi/2)(x + \xi) + \dot{\theta}(t)^2(x + \xi) + 2\dot{\theta}(t)R(-\pi/2)\dot{\xi}.
$$

Fig. 6.5 The elastic body, its boundary components Γ_C, Γ_D, and the contact set

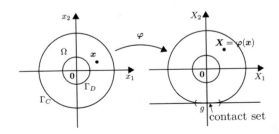

In the above, the stress tensor has been denoted as

$$\sigma[\xi] = 2\mu\epsilon[\xi] + \lambda(\text{div}\xi)I,$$

where μ, λ are the Lamé coefficients, and the strain tensor is

$$\epsilon[\xi] = \frac{1}{2}\left(\nabla\xi + \nabla\xi^T\right).$$

The matrix R specifies a counter-clockwise rotation by the angle θ, I is the identity map, and e_1, e_2 are the unit basis vectors.

The unknown contact set is

$$\{\varphi(x); \quad x \in \Gamma_C, \ \varphi_2(x) = g\},$$

where $\varphi(x, t) = R(\theta(t))(x + \xi(x, t))$, for $x \in \overline{\Omega}$ and all $t \geq 0$.

Given ξ^0 and setting a small time step $h > 0$, we define $\xi^{-1} = \xi^0 - h\eta^0$. Then the DMF is able to approximate the solution of the model equation by minimizing the following functionals for $k = 1, 2, \ldots, M$:

$$\mathcal{J}^k(\xi) = \rho \int_\Omega \frac{|\xi - 2\xi^{k-1} + \xi^{k-2}|^2}{2h^2} \, dx + \frac{1}{2}\int_\Omega \left(\frac{1}{2}\sigma[\xi] + \sigma[\xi^{k-2}]\right) : \epsilon[\xi]\, dx$$

$$- \rho \int_\Omega f^{k-1} \cdot \xi dx \qquad (6.15)$$

within the admissible set:

$$\mathcal{K}^k = \left\{\xi \in W^{1,2}(\Omega; \mathbb{R}^2); \ \xi = 0 \text{ a.e. on } \Gamma_D, (I + \xi) \cdot \left(R^T(\theta^k)e_2\right) \geq g^k \text{a.e. on } \Gamma_C\right\}.$$

Here $\sigma : \epsilon = \sigma_{ij}\epsilon_{ij}$ is the inner product of the two second-order tensors. Also, since time has been discretized, we have defined the following approximations:

$$f^k(x) = f(kh, x, \xi^k, (\xi^k - \xi^{k-1})/h),$$

$$\theta^k = \theta(kh),$$

$$g^k = g(kh).$$

Note that the energy-preserving Crank–Nicolson scheme is employed in the second term, which turns out to be essential to get physically reasonable results.

Numerical results for two types of deformation of the annular body $\Omega = \{x \in \mathbb{R}^2; \ r_D < |x| < r_C\}$ with $r_D = 0.25$, $r_C = 0.5$ are displayed in Figs. 6.6 and 6.7, respectively. The first result is for a simple compression test where the obstacle is

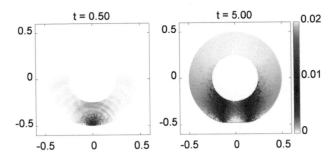

Fig. 6.6 Compression of an elastic body against an obstacle (no rotation). The magnitude of the stress is displayed

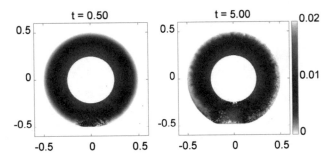

Fig. 6.7 Contact dynamics between an obstacle and an elastic body, rotating counter-clockwise with angular velocity $\omega = \pi/10$. The magnitude of the stress is displayed

prescribed to move as

$$g(t) = \min\left\{0.005t, \frac{r_C - r_D}{10}\right\} - r_C,$$

and the remaining parameters are chosen as in [2]. In the second simulation, in addition the roller is rotating with angular velocity $\omega = \pi/10$.

6.2.3 Elastic Shell Impact

Here we will show the application of the DMF to solving a model equation describing the impact of an elastic closed shell with a flat obstacle. We remark that although reduced problems such as the elastic bounce of a mass point [74–76] were analyzed, the problem of a shell is too complex for a rigorous analysis. On the other hand, the vector-valued DMF model of a bouncing shell described below is quite simple and allows for an ingenuous numerical implementation.

The elastic shell is described by a closed curve $\boldsymbol{p}(t, \theta) = (p(t, \theta), q(t, \theta))$, where $\theta \in [0, 2\pi]$ is the parameterization variable, and the outer unit normal and tangent vectors are (subscripts denote partial differentiation):

$$v(t, \theta) = \frac{1}{|p_\theta|}(q_\theta(t, \theta), -p_\theta(t, \theta)), \qquad \tau(t, \theta) = \frac{p_\theta(t, \theta)}{|p_\theta(t, \theta)|}.$$

A dimensionless model equation is the following hyperbolic free boundary problem (see [32] for the derivation):

$$\chi_{\{q>0\}} p_{tt} = \left(-|p_\theta| \left(\kappa_{ss} + \frac{1}{2}\kappa^3 \right) \chi_{\{q>0\}} + \frac{1}{2}|p_\theta|\kappa - a_1|p_\theta| \left(|p_\theta| - 1 \right) \kappa \chi_{\{q>0\}} \right) v$$

$$+ a_2|p_\theta| \left(\frac{1}{V} - \frac{1}{\pi} \right) \chi_{\{q>0\}} v + a_1|p_\theta||p_\theta|_s \chi_{\{q>0\}}\tau + a_3\chi_{\{q>0\}}e_2,$$

where

$$a_1 = 12\frac{r_0^2}{d^2}, \qquad a_2 = 12\frac{r_0 c_g}{kd^3}, \qquad \text{and} \qquad a_3 = 12\frac{g\sigma r_0^3}{kd^2}.$$

Here κ is the curvature of the shell, $d > 0$ is a parameter describing the thickness of the shell, r_0 is the natural radius of the circular shell, σ is the mass density of the shell in equilibrium, k is an elastic coefficient, e_2 is the unit vector in the direction of the y-axis, V is the time-dependent volume of the shell, g is a gravitational constant, and c_g is constant related to the total mass of the enclosed gas and the speed of sound.

As the problem has variational structure, the DMF approach is applicable to get approximating solutions of the model equation. We discretize time as $h = T/N$, and consider the successive constrained minimization of the functionals

$$J_n(p) = \frac{1}{2}d\sigma r_0 \int_0^{2\pi} \frac{p \cdot (\frac{1}{2}p - 2p^{n-1} + p^{n-2})}{h^2}\chi_{\{q>0\}}\, d\theta + E_e(p) + E_p(p) + E_g(p) \tag{6.16}$$

in the set

$$\mathcal{K} = \left\{ p \in \left[H^2_{per}((0, 2\pi)) \right]^2; \ q \ge 0 \right\}$$

of periodic H^2-functions with nonnegative second component. In particular, p^n is given as the minimizer in $\mathcal{K}$ of J_n for $n = 2, 3, \ldots, N$. Here, the elastic energy is defined as

$$E_e(p) = \frac{1}{24}kd^3 \int_0^{2\pi} (\kappa - \kappa_0)^2|p_\theta|\, d\theta + \frac{1}{2}kd \int_0^{2\pi} \left(\frac{|p_\theta|}{|q_\theta|} - 1 \right)^2 |q_\theta|\, d\theta,$$

where $\kappa_0 > 0$ is the bending energy reference, and the equilibrium shape of the shell (not in contact with the obstacle) is denoted by $q(\theta)$. Furthermore, the potential energy is defined as

$$E_p(\boldsymbol{p}) = gd \int_0^{2\pi} \sigma |\boldsymbol{q}_\theta| q \, \chi_{\{q>0\}} \, d\theta,$$

and the energy due to compression of the enclosed gas is modeled by

$$E_g(\boldsymbol{p}) = -P_0(V - V_0) - c_g \left(1 - \frac{V}{V_0} + \ln \frac{V}{V_0} \right),$$

where P_0 is the uniform pressure of the gas in the equilibrium shell and V_0 is its volume.

The discrete Morse flow approach can also naturally handle problems with a prescribed volume:

$$\frac{1}{2} \int_0^{2\pi} \boldsymbol{p} \cdot \begin{pmatrix} 0 & 1 \\ -1 & 0 \end{pmatrix} \boldsymbol{p}_\theta \, d\theta = V_0,$$

by including the constraint in the set $\mathcal{K}$ and, in numerical implementation, in the functional J_n via an additional penalty term.

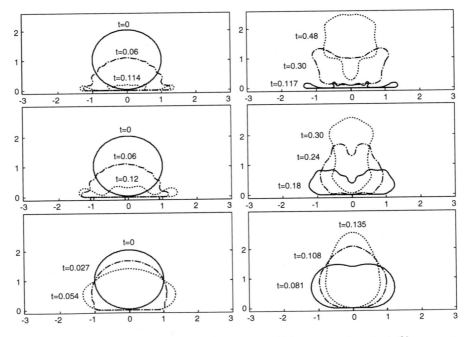

Fig. 6.8 Collision dynamics for different values of a_2, which expresses the extent of incompressibility of the enclosed gas (by rows, $a_2 = 0, 384$ and $98,304$)

An example of simulation of a bouncing ring is shown in Fig. 6.8. The initial functions for the DMF are chosen as $p^0 = q = (r_0 \cos \theta, r_0 \sin \theta + 1)$ and $p^1 = p^0 - (0, 16h)$. Parameters are set corresponding to $a_1 = 15{,}000$ and $a_3 = 0$, while a_2 takes the values 0, 384 and 98,304, where larger values of a_2 mean stricter demands on volume preservation.

References

1. Aguilera, N., Alt, H., Caffarelli, L.A.: An optimization problem with volume constraint. SIAM J. Control Optim. **24**, 191–198 (1986)
2. Akagawa, Y., Morikawa, S., Omata, S.: A numerical approach based on variational methods to an elastodynamic contact problem. Sci. Rep. Kanazawa Univ. **63**, 29–44 (2019)
3. Akagawa, Y., Ginder, E., Koide, S., Omata, S., Svadlenka, K.: A Crank–Nicolson type minimization scheme for a hyperbolic free boundary problem. Discrete Contin. Dynam. Syst. Ser. B **27**(5), 2661–2681 (2022)
4. Alt, H.W., Caffarelli, L.A.: Existence and regularity for a minimum problem with free boundary. J. Reine Angew. Math. **325**, 105–144 (1981)
5. Alt, H.W., Caffarelli, L.A., Friedman, A.: A free boundary problem for quasi-linear elliptic equations. Ann. Scuola Norm. Sup. Pisa Cl. Sci. **11**(4), 1–44 (1984)
6. Amerio, L.: A unilateral problem for a non linear vibrating string equation. Atti della Accademia Nazionale dei Lincei. Rendiconti Serie **8**, 64(1), 8–21 (1978)
7. Bamberger, A., Schatzman, M.: New results on the vibrating string with a continuous obstacle. SIAM J. Math. Anal. **14**(3), 560–595 (1983)
8. Bellettini, G., Novaga, M., Orlandi, G.: Time-like minimal submanifolds as singular limits of nonlinear wave equations. Physica D **239**, 335–339 (2010)
9. Bellettini, G., Novaga, M., Orlandi, G.: Lorentzian varifolds and applications to relativistic strings. Indiana Univ. Math. J. **61**(6), 2251–2310 (2012)
10. Berestycki, H., Cafafrelli, L.A., Nirenberg, L.: Uniform estimates for regularization of free boundary problems. In: Analysis and Partial Differential Equations. Marcel Dekker, New York (1990)
11. Bonafini, M., Novaga, M., Orlandi, G.: A variational scheme for hyperbolic obstacle problems. Nonlin. Anal. **188**, 389–404 (2019)
12. Bonafini, M., Le, V.P.C., Novaga, M., Orlandi, G.: On the obstacle problem for fractional semilinear wave equations. Nonlin. Anal. **210**, 112368 (2021)
13. Burridge, R., Keller, J.B.: Peeling, slipping and cracking – some one-dimensional free-boundary problems in mechanics. SIAM Rev. **20**(1), 31–61 (1978)
14. Caffarelli, L.A., Vázquez, J.L.: A free-boundary problem for the heat equation arising in flame propagation. Trans. Am. Math. Soc. **347**(2), 411–441 (1995)
15. Coclite, G.M., Florio, G., Ligabo, M., Maddalena, F.: Nonlinear waves in adhesive strings. SIAM J. Appl. Math. **77**(2), 347–360 (2017)
16. Coclite, G.M., Florio, G., Ligabo, M., Maddalena, F.: Adhesion and debonding in a model of elastic string. Comput. Math. Appl. **78**(6), 1897–1909 (2019)

© The Author(s), under exclusive license to Springer Nature Singapore Pte Ltd. 2022
S. Omata et al., *Variational Approach to Hyperbolic Free Boundary Problems*,
SpringerBriefs in Mathematics, https://doi.org/10.1007/978-981-19-6731-3

17. Coclite, G.M., Devillanova, G., Maddalena, F.: Waves in flexural beams with nonlinear adhesive interaction. Milan J. Math. **89**, 329–344 (2021)
18. Dal Maso, G., Lazzaroni, G., Nardini, L.: Existence and uniqueness of dynamic evolutions for a peeling test in dimension one. J. Differential Equations **261**, 4897–4923 (2016)
19. Dal Maso, G., De Luca, L.: A minimization approach to the wave equation on time-dependent domains. Adv. Calc. Var. **13**(4), 425–436 (2020)
20. de Gennes, P.-G., Brochard-Wyart, F., Quéré, D.: Capillarity and Wetting Phenomena: Drops, Bubbles, Pearls, Waves. Springer (2004)
21. del Pino, M., Jerrard, R.L., Musso, M.: Interface dynamics in semilinear wave equations. Commun. Math. Phys. **373**, 971–1009 (2020)
22. El Smaily, M., Jerrard, R.L.: A refined description of evolving interfaces in certain nonlinear wave equations. Nonlinear Differential Equations Appl. **25**, article no. 15 (2018)
23. Frontini, M., Gotusso, L.: Numerical study of the motion of a string vibrating against an obstacle by physical discretization. Appl. Math. Model. **14**, 489–494 (1990)
24. Giaquinta, M.: Multiple Integrals in the Calculus of Variations and Nonlinear Elliptic Systems. Princeton University Press (1983)
25. Gilbarg, D., Trudinger, N.S.: Elliptic Partial Differential Equations of Second Order. Springer, Berlin (1983)
26. Ginder, E., Svadlenka, K.: A variational approach to a constrained hyperbolic free boundary problem. Nonlin. Anal. Theory Methods Appl. **71**(12), e1527–e1537 (2009)
27. Ginder, E., Svadlenka, K.: Wave-type threshold dynamics and the hyperbolic mean curvature flow. J. Ind. Appl. Math. **33**(2), 501–523 (2016)
28. Imai, H., Kikuchi, K., Nakane, K., Omata, S., Tachikawa, T.: A numerical approach to the asymptotic behavior of solutions of a one-dimensional hyperbolic free boundary problem. JJIAM **18**(1), 43–58 (2001)
29. Ishida, S., Yamamoto, M., Ando, R., Hachisuka, T.: A hyperbolic geometric flow for evolving films and foams. ACM Trans. Graph. **36**(6), article no. 199, 1–11 (2017)
30. Jerrard, R.L.: Defects in semilinear wave equations and timelike minimal surfaces in Minkowski space. Anal. PDE **4**(2), 285–340 (2011)
31. Kazama, M., Omata, S.: Modeling and computation of fluid–membrane interaction. Nonlin. Anal. Theory Methods Appl. **71**(12), e1553–e1559 (2009)
32. Kazama, M., Kikuta, A., Nagasawa, T., Omata, S., Svadlenka, K.: A global model for impact of elastic shells and its numerical implementation. Adv. Math. Sci. Appl. **23**(1), 93–108 (2013)
33. Kikuchi, N.: An approach to the construction of Morse flows for variational functionals. In: Coron, J.-M., Ghidaglia, J.-M., Hélein, F. (eds.) Nematics – Mathematical and Physical Aspects, NATO Adv. Sci. Inst. Ser. C: Math. Phys. Sci., vol. 332, pp. 195–198. Kluwer Acad. Publ., Dodrecht (1991)
34. Kikuchi, K.: Constructing a solution in time semidiscretization method to an equation of vibrating string with an obstacle. Nonlin. Anal. **71**, e1227–e1232 (2009)
35. Kikuchi, K., Omata, S.: A free boundary problem for a one dimensional hyperbolic equation. Adv. Math. Sci. Appl. **9**(2), 775–786 (1999)
36. Kim, J.U.: A boundary thin obstacle problem for a wave equation. Commun. Partial Differential Equations **14**(8–9), 1011–1026 (1989)
37. Koshizuka, S., Oka, Y.: Moving-particle semi-Implicit method for fragmentation of incompressible fluid. Nuclear Sci. Eng. **123**, 421–434 (1996)
38. Ladyzhenskaya, O., Uraltseva, N.: Linear and Quasilinear Elliptic Equations. Academic Press, New York (1968)
39. Lazzaroni, G., Nardini, L.: On the quasistatic limit of dynamic evolutions for a peeling test in dimension one. J. Nonlinear Sci. **28**, 269–304 (2018)
40. Lazzaroni, G., Nardini, L.: Analysis of a dynamic peeling test with speed-dependent toughness. SIAM J. Appl. Math. **78**, 1206–1227 (2018)
41. Lazzaroni, G., Nardini, L.: On the 1d wave equation in time-dependent domains and the problem of debond initiation. ESAIM: COCV **25**, 80 (2019)

42. Leonardi, G., Tilli, P.: On a constrained variational problem in the vector-valued case. J. Math. Pures Appl. **85**, 251–268 (2006)
43. Li, T.: Global Classical Solutions for Quasilinear Hyperbolic Systems. Research in Applied Mathematics, vol. 32. Masson/John Wiley (1994)
44. Li, T., Yu, W.: Some existence theorems for quasilinear hyperbolic systems of partial differential equations in two independent variables, I: Typical boundary value problems. Sci. Sin. **13**, 529–549 (1964)
45. Li, T., Yu, W.: Some existence theorems for quasilinear hyperbolic systems of partial differential equations in two independent variables, II: Typical boundary value problems of functional form and typical free boundary problems. Sci. Sin. **13**, 551–562 (1964)
46. Li, T., Yu, W.: Boundary Value Problems for Quasilinear Hyperbolic Systems. Duke University Mathematics Series V (1985)
47. Mabrouk, M.: A unified variational model for the dynamics of perfect unilateral constraints. Eur. J. Mech. A/Solids **17**, 819–842 (1998)
48. Majda, A.J.: The stability of multi-dimensional shock fronts. Mem. Am. Math. Soc. **41**(275) (1983)
49. Majda, A.J., The existence of multi-dimensional shock fronts. Mem. Am. Math. Soc. **43**(281) (1983)
50. Majda, A.J.: Compressible Fluid Flow and Systems of Conservation Laws in Several Space Variables. Springer, (2012)
51. Majda, A.J., Souganidis, P.E.: Existence and uniqueness of weak solutions for precipitation fronts: A novel hyperbolic free boundary problem in several space variables. Commun. Pure Appl. Math. **63**(10), 1351–1361 (2010)
52. Maruo, K.: Existence of solutions of some nonlinear wave equation. Osaka J. Math. **22**, 21–30 (1985)
53. Maruo, K.: On certain nonlinear differential equations of second order in time. Osaka J. Math. **23**, 1–53 (1986)
54. Mohammad, R.Z., Svadlenka, K.: Multiphase volume-preserving interface motions via localized signed distance vector scheme. Discrete Contin. Dynam. Syst. Ser. S **8**(1), 969–988 (2015)
55. Monaghan, J.J.: Simulating free surface flows with SPH. J. Comput. Phys. **110**, 399–406 (1994)
56. Morrey, Jr., C.B: Multiple Integrals in the Calculus of Variations. Springer, Berlin (1966)
57. Nagai, K., Tachibana, K., Tobe, Y., Kazama, M., Kitahata, H., Omata, S., Nagayama, M.: Mathematical model for self-propelled droplets driven by interfacial tension. J. Chem. Phys. **144**, 114707 (2016)
58. Nagasawa, T., Omata, S.: Discrete Morse semiflows of a functional with free boundary. Adv. Math. Sci. Appl. **2**(1), 147–187 (1993)
59. Nagasawa, T., Tachikawa, A.: Existence and asymptotic behavior of weak solutions to strongly damped semilinear hyperbolic systems. Hokkaido Math. J. **24**(2), 387–405 (1995)
60. Nagasawa, T., Nakane, K., Omata, S.: Numerical computations for motion of vortices governed by a hyperbolic Ginzburg-Landau system. Nonlinear Anal. Ser A Theory Methods **51**(1), 67–77 (2002)
61. Nakane, K., Shinohara, T.: Global solutions to a one-dimensional hyperbolic free boundary problem which arises in peeling phenomena. J. Comput. Appl. Math. **152**, 367–375 (2003)
62. Nakane, K., Shinohara, T.: Existense of a periodic solution for a free boundary problem of hyperbolic type. J. Hyperbolic Differential Equations **5**(4), 785–806 (2008)
63. Nocedal, J., Wright, S.J.: Numerical Optimization. Springer (2006)
64. Omata, S.: A free boundary problem for a quasilinear elliptic equation, Part I: Rectifiability of free boundary. Differential Integral Equations **6**(6), 1299–1312 (1993)
65. Omata, S.: A numerical method based on the discrete morse semiflow related to parabolic and hyperbolic equation. Nonlinear Anal. Theory Methods Appl. **30**(4), 2181–2187 (1997)
66. Omata, S.: A numerical treatment of film motion with free boundary. Adv. Math. Sci. Appl. **14**, 129–137 (2004)
67. Omata, S.: A hyperbolic free boundary problem and its numerical and mathematical analysis. Sugaku Expositions **33**(1), 31–55 (2020)

68. Omata, S., Yamaura, Y.: A free boundary problem for quasilinear elliptic equations. Proc. Japan Acad. Ser. A Math. **21**, 281–286 (1990)

69. Omata, S., Yamaura, Y.: A free boundary problem for quasilinear elliptic equations, Part II: $C^{1,\alpha}$-regularity of free boundary. Funkcialaj Ekvacioj **42**(1), 9–70 (1999)

70. Omata, S., Kazama, M., Nakagawa, H.: Variational approach to evolutionary free boundary problems. Nonlinear Anal. Theory Methods Appl. **71**(12), e1547–e1552 (2009)

71. Panet, M., Paoli, L., Schatzman, M.: Theoretical and numerical study for a model of vibrations with unilateral constraints. In: Raous, M., Jean, M., Moreau, J.J. (eds.) Contact Mechanics. Springer, Boston (1995)

72. Paoli, L., Schatzman, M.: A numerical scheme for impact problems I: the one-dimensional case. SIAM J. Numer. Anal. **40**(2), 702–733 (2002)

73. Paoli, L., Schatzman, M.: A numerical scheme for impact problems II: the multidimensional case. SIAM J. Numer. Anal. **40**(2), 734–768 (2003)

74. Percivale, D.: Uniqueness in the elastic bounce problem. J. Differential Equations **56**(2), 206–215 (1985)

75. Percivale, D.: Bounce problem with weak hypotheses of regularity. Ann. Mat. Pura Appl. **143**, 259–274 (1986)

76. Percivale, D.: Uniqueness in the Elastic Bounce Problem II. J. Differential Equations **90**, 304–315 (1991)

77. Rothe, E.: Zweidimensionale parabolische Randwertaufgaben als Grenzfall eindimensionaler Randwertaufgaben. Math. Ann. **102**, 650–670 (1930)

78. Schatzman, M.: A class of nonlinear differential equations of second order in time. Nonlinear Anal. **2**(3), 355–373 (1978)

79. Schatzman, M.: A hyperbolic problem of second order with unilateral constraints: the vibrating string with a concave obstacle. J. Math. Anal. Appl. **73**, 138–191 (1980)

80. Schatzman, M.: The penalty method for the vibrating string with an obstacle. In: Analytical and Numerical Approaches to Asymptotic Problems in Analysis (Proc. Conf., Univ. Nijmegen, Nijmegen, 1980), volume 47 of North-Holland Math. Stud., pp. 345–357. North-Holland, Amsterdam (1981)

81. Serra, E., Tilli, P.: Nonlinear wave equations as limits of convex minimization problems: proof of a conjecture by De Giorgi. Ann. Math. **175**, 1551–1574 (2012)

82. Serra, E., Tilli, P.: A minimization approach to hyperbolic Cauchy problems. J. Eur. Math. Soc. **18**, 2019–2044 (2016)

83. Svadlenka, K.: Mathematical analysis and numerical computation of volume-constrained evolutionary problems, involving free boundaries. PhD thesis, Kanazawa University, 2008

84. Svadlenka, K., Omata, S.: Mathematical modelling of surface vibration with volume constraint and its analysis. Nonlinear Anal. **69**(9), 3202–3212 (2008)

85. Svadlenka, K., Omata, S.: Mathematical analysis of a constrained parabolic free boundary problem describing droplet motion on a surface. Indiana Univ. Math. J. **58**(5), 2073–2102 (2009)

86. Tachikawa, A.: A variational approach to constructing weak solutions of semilinear hyperbolic systems. Adv. Math. Sci. Appl. **4**, 93–103 (1994)

87. Taguti, T.: Dynamics of simple string subject to unilateral constraint: A model analysis of sawari mechanism. Acoust. Sci. Technol. **29**(3), 203–214 (2008)

88. Tilli, P.: On a constrained variational problem with an arbitrary number of free boundaries. Interfaces Free Bound **2**, 201–212 (2000)

89. Yamaura, Y.: The regularity of minimizers of a radially symmetric free boundary problem. Ann. Univ. Ferrara Sez. VII (38), 177–192 (1992)

90. Yamazaki, T., Omata, S., Svadlenka, K., Ohara, K.: Construction of approximate solution to a hyperbolic free boundary problem with volume constraint and its numerical computation. Adv. Math. Sci. Appl. **16**(1), 57–67 (2006)

91. Yazaki, S.: On the tangential velocity arising in a crystalline approximation of evolving plane curves. Kybernetika **43**(6), 913–918 (2007)

92. Yoshiuchi, H., Omata, S., Svadlenka, K., Ohara, K.: Numerical solution of film vibration with obstacle. Adv. Math. Sci. Appl. **16**(1), 33–43 (2006)

Printed in the United States
by Baker & Taylor Publisher Services